Structural Design & Drawing

Structural Design & Drawing

(STEEL STRUCTURES)

VOLUME-3

Dr. D. KRISHNAMURTHY
Ph.D. (Glas), ME (Cal) BE (Hons),
MIE (Inc). MACI, MIABSE (Eng)
Professor of Civil Engineering
M. A. College of Technology, Bhopal
(Presently Professor of Civil Engineering
Sulemaniah University
Sulemaniah)

CBS

CBS PUBLISHERS & DISTRIBUTORS PVT. LTD.

New Delhi • Bengaluru • Chennai • Kochi • Mumbai • Pune

Structural Design & Drawing

(Steel Structures)

ISBN: 978-81-239-0148-0

First Edition: 1995

Reprint: 1992, 1999, 2003, 2006, 2008, 2009, 2010, 2011, 2012, 2014, 2015, **2024**

Published by **Satish Kumar Jain** and produced by **Varun Jain** for

CBS Publishers & Distributors Pvt Ltd

4819/XI Prahlad Street, 24 Ansari Road, Daryaganj, New Delhi 110 002, India.

Ph: 011-23289259, 23266861 Website: www.cbspd.com

e-mail: delhi@cbspd.com

Corporate Office: 204 FIE, Industrial Area, Patparganj, Delhi 110 092

Ph: 011-4934 4934 Fax: 011-4934 4935

e-mail: publishing@cbspd.com; publicity@cbspd.com

Branches

- **Bengaluru:** Seema House 2975, 17th Cross, KR Road, Banasankari 2nd Stage, Bengaluru 560 070, Karnataka, India
  Ph: +91-80-26771678/79 Fax: +91-80-26771680 e-mail: bangalore@cbspd.com
- **Chennai:** 7, Subbaraya Street, Shenoy Nagar, Chennai 600 030, Tamil Nadu, India
  Ph: +91-44-26680620, 26681266 Fax: +91-44-42032115 e-mail: chennai@cbspd.com
- **Kochi:** 42/1325, 1326, Power House Road, Opp KSEB, Power House, Ernakulum Kochi 682 018, Kerala, India
  Ph: +91-484-4059061-65,67 Fax: +91-484-4059065 e-mail: kochi@cbspd.com
- **Kolkata:** 147, Hind Ceramics Compound, 1st Floor, Nilgunj Road, Belghoria, Kolkata-700056, West Bengal, India
  Ph: +033-25633055, 033-25633056 e-mail: kolkata@cbspd.com
- **Lucknow:** Basement, Khushnuma Complex, 7 Meerabai Marg (Behind Jawahar Bhawan), Lucknow-226001, UP, India
  Ph: +0522-4000032 e-mail: tiwari.lucknow@cbspd.com
- **Mumbai:** PWD Shed, Gala no 25/26, Ramchandra Bhatt Marg, Next to JJ Hospital Gate no. 2, Opp. Union Bank of India, Noorbaug, Mumbai-400009, Maharashtra, India
  Ph: 022-66661880/89 e-mail: mumbai@cbspd.com

Representatives

- Hyderabad 0-9885175004
- Jharkhand 0-9811541605
- Nagpur 0-8692091830
- Patna 0-9334159340
- Pune 0-9664372571
- Uttarakhand 0-9716462459

Printed at Rashtriya Printers, Dilshad Garden, Delhi, India

Preface

Structural engineering is a branch of civil engineering which has to do with the dcsign of buildings, bridges, water towers retaining walls, silos, bunkers chimneys and other types of civil engineering structures.

Structural design of any structure may be divided into :

(*i*) Assessment and distribution of loads which the structural component is required to support or subjected to.

(*ii*) Proportioning the trial section of the components or the appropriate arrangement of structural elements.

(*iii*) Computation of bending moments, shear forces and direct forces due to critical combination of loads to which the structural member is subjected or is required to carry.

(*iv*) Selection of proper size and shape of the member to resist the forces safely keeping in view, the economy in the design.

(*v*) Preparation of lay out of the structure and the finished working drawings with all dimensions for all members which will be required for executing the construction of the structure at site or for fabrication of the structure in the factory.

Students taking courses on structural design have usually no great difficulty in computations involved in the structural design but they often find that the available text books afford very little guidance in the actual sequence in the design, appropriate choice of sections on the basis of strength and economy and the detailed dimensioned drawings.

The author has endeavoured to provide these needs to the students, structural designers handling designs using concrete and steel as materials and also to provide a detailed guide for young engineers and architects engaged in consultancy services.

In the process of making a complete design, special attention has been devoted to giving full details of each step. Each design is provided with fully dimensioned working drawings. Designs of all kinds of practical problems that the designer normally comes across are presented using a standard sequence, along with the fully dlmensioned drawings in the three volumes.

Volume I. Elementary Structural Design and Drawing is in two parts. Part I deals with the designs of R.C.C. Slabs, beams columns footings and stair cases, basically design and drawings of all basic components of buildings. Part II deals with the designs of structural steel connections pin, riveted, bolted and welded connectlons—beam stanchion, grillage foundations, roof trusses and riveted plate girders —again all components of residential and office buildings and industrial sheds with steel as structural material. This volume has been designed by the author for the greatest usefulness to the students of diploma, pre-final year civil engineering degree and to those appearing for section 'A' of professional examinations in addition to these practising engineers and architects involved in design buildings.

Volume II. Structural Design and Drawing (R.C.C. Structures) deals with complete designs with detailed drawings fully dimension on practical problems on retaining walls, domed, roofs, water tanks on ground, underground, overground reserviors. silos. bunkers, slab and T-beam bridges. This book is intended to serve as a text book to civil engineering students of finai year, degree course and to those appearing for Section 'B' of the professional examination and as a reference book for practising engineers in public works, public health departments and stuctural designers in consulting firms.

Volume III. Structural Design and Drawing (Steel Structures) deals with complete designs with drawings railway and highway, welded plate girders, water tanks, guyed and self supporting chimneys. This volume serves the needs of final year engineering degree students and those appearing for Section 'B' of the professional examinations as a text book and as a reference book to those engaged in structural designs in steel. Railway engineers, hlghway engineers and structural engineers in firms dealing with industrial buildings and bridges etc. will find it of great use.

AUTHOR

Contents

PART I

WELDED PLATE GIRDERS

1. Welded Connections

1·1. Introduction

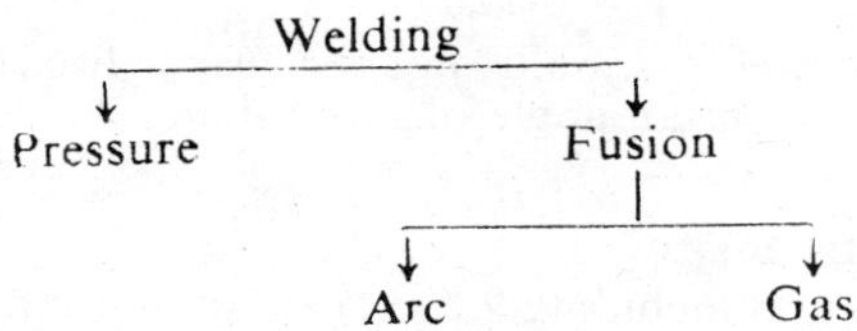

1·2. Advantages: It is a modern method.

Economy—Elimination of connecting angles, plates, rivet heads, guessets, splice plates and labour.

Stiffness—Connections are rigid; Efficiency—100% can be achieved; Others—Noiseless operation, clean and neat joints. Dead load reduction.

1·3. Disadvantages:

Over rigidity—warping;
Fatigue
Brittle
x-ray testing and inspection—skilled labour

1·4. Types

Butt welds—Single *V*, *U*, *J*
Fillet welds—Double *V*, *U*, *J*

1·5. Definitions

β = Angle of Bevel
α = Included angle

Fillet welds—A weld of approximately triangular cross-section joining two surface at right angles to each other in a Lap, tee or corner joint.

Leg—Fusion face in a fillet weld.

Leg length—The distance from the root to the toe of a fillet weld, measured along the fusion face.

1·6. Size:

The size of the weld shall be specified by its effective throat thickness.

1·7. Effective throat thickness:

In the case of complete penetration on butt weld, the effective throat thickness shall be taken as the thickness of the thinner part joined.

In the case of a fillet weld, the effective throat thickness shall be taken as :

K times fillet size where K is constant which depends on the angle between the fusion faces.

1·8.

Angle	K
60°—90°	0·7
91°—100°	0·65
101°—106°	0·60
107°—113°	0·55
114°—120°	0·5

FIllet welds should not be used for connecting parts whose fusion faces form an angle of more than 120° or less than 60°.

(∢ 60° ⊁ 120°)

1·9. Minimum size

Upto and including 9·5 mm	3 mm size
over 9·5 mm including 19 mm	5 mm
Over 19 mm—32 mm	6 mm
Over 32 mm—50 mm	8 mm
Over 50 mm	Special precautions are necessary.

1·10. Permissible stresess (Butt welds)

Direct tension	1420 kg/cm²
,, comp	1420 kg/cm²

In bending

Tension	1575 kg/cm²
Comp	1575 kg/cm²
In shear	1025 kg/cm²

Site welds 80% of above stress

Increase in stressess

When wind loads are considered—an increase of 25% is allowed.

1·11. Minimum length

The effective length of a fillet weld designed to transmit loading shall not be less than four times the size of the weld or ∢ 3·8 cm.

1·12. Spacing ⊁ 30 cm

or 16 × thickness of the thinner part in compression
or 24 × thickness of the thinner part in tension
End returns = 2 × size of the weld
Size of the weld = S
Effective size = 0·7S
Effective length = l
Allowable stress in shear = 1025 kg/cm²
Strength of weld = 0·75 × l × 1025.

1·13. Welding Symbols

- Bead
- Fillet
- Square
- V
- Bevel
- U
- J
- Plug or slot
- Field weld
- Weld arongd
- Flush

2. Plate Girders—General Information

2·1. Introduction

Plate girders are used when standard rolled sections do not provide the required section modulus. Building up a beam by welding or rivetting plates and angles to increase the modulus of section is one of the choices available to the designer.

Plate girders are especially adopted for short spans and heavy loads.

2·2. Types

(*i*) Rivetted plate girders (*ii*) Welded plate girders.

These girders can be used for buildings, highway bridges and Railway bridges.

2·3. Components

In a rivetted plate girder the components are

(*i*) Web plate (*ii*) Flange plates (*iii*) Flange angles (*iv*) Stiffness: (*a*) Intermediate stiffners (*b*) Bearing stiffners: (1) Under reactions (2) Under concentrated loads (*c*) Longitudinal stiffners (*v*) Splicing.

The two main components of the plate girder namely flanges and web are designed to take bending moment and shear force respectively and therefore section of the plate girder can be proportioned by varying the thickness of flange and depth of the web to take up varying bending moment and shearing force at any section.

However, in the case of a welded plate girder, the components namely flange angles and splicing are not necessary.

2·4. Effective Span

The effective span for a plate girder is usually the centre to centre of end bearings.

Clear span *Metres*	*Effective span* *Metres*
3	3·300
6	6·450
9	9·520
12	12·600
15	15·675
18	18·750
21	21·825
24	24·900

2·5. Proportioning of components

The approximate depth of the girder is usually taken as 1/8 to 1/12 of the effective span L.

2·5·1. Depth of Girder

For higher loads—$\frac{L}{12}$

For midium loads—$\frac{L}{10}$

For heavy loads—$\frac{L}{8}$

The economical depth of the girder can be obtained from the expression

$$d=K\sqrt[3]{M/f_b}$$

Where, M=Maximum bending moment

f_b=Maximum bending stress permissible for a laterally supported girder=1500 kg/cm^2

K=Parameter=5 for welded girder=5.5 for rivetted girder.

2·5·2. Width of Girder

The width of the girder should be within the limits of $L/40$ to $L/45$. The maximum width given by this rule may be exceeded provided the flange plates are adequately stiffened.

2·6. Minimum self weight

To obtain the minimum weight of steel in plate girder design, several different depths should be used in a variety of preliminary designs to determine the trend of weight with respect to variations in plate girder depth.

If longitudinal stiffners are used, the web depth-thickness ratio may be increased above 200, but in short span plate girders used in building construction use of longitudinal stiffners introduces considerable complexity in the framing.

If a very long span girder of 30 m or more is required, then the possibility of economy through the use of longitudinal stiffners should be investigated. Such stiffners are commonly used in continuous span highway bridge girders.

The self weight of welded plate girders can be estimated from the expressions

(*i*) $w=3M/f_bd$, where w=weight in kg/m, M=Bending moment in cm kg, f_b=Allowable bending stress in kg/cm^2: d=Depth of girder in cm.

(*ii*) $w=W/300$ where, W=total superimposed load on the girder in kg, and w=Weight in kg/m.

The weight can also be estimated at within close enough design limits by estimating the weights of flange plates with angles if used and web plates and adding the following percentages for weights of stiffners and other details.

Welded girders—30% of web weight. Rivetted girders with cripped stiffners—50% of web weight

Rivetted girders with filler plates under all stiffners—70% of web weight.

2·7. Limiting Deflection

The maximum deflection should not exceed 1/325 of the span, but this limit may be exceeded in cases where greater deflection would not impair the strength or efficiency of the structure or lead to damage to finishings.

2·8. Allowable Stresses

Allowable bending stress for plate girders wlth single or multiple webs upto and including 20 mm thick = 1575 kg/cm^2

Over 20 mm thick = 1500 kg/cm^2

Allowable shear stress for plate Girders with webs upto and including 20 mm thick = 945 kg/cm^2

For over 20 mm thick = 865 kg/cm^2

2·9. Minimum Thickness of Plates

For those exposed to wheather but accessible for painting = 6mm

For those exposed to weather but not accessible for painting = 8mm

For highway bridges = 8 mm

For Railway bridges = 10 mm

For welded crane gantry plate girders intended for carrying cranes with a lifting load of 15 tonnes or more = 8 mm

2·10. Analysis

Plate girders are proportioned on the basis of the moment of inertia of the gross cross-section with neutral axis taken at the centroid of the section. The maximum compression and tensile stresses in bending shall be calculated by increasing the stresses computed on the basis of the gross moment of inertia by the ratio of gross area to net area of the compressive and tensile flange sections respectively. The flange section for this purpose shall be taken in riveted construction to be the flange plate, flange angles and that part of the web which is between the flange angles, and in welded construction to be the flange plates plus that of the tongue plates, (if any) upto a limit of eight times their thickness which shall be not less than twice the thickness of the web. The maximum stresses so obtained, however, shall not exceed the specified allowable stresses.

2·11. Riveted Girder

Plate girder is a built-up beam and the moment of resistance can be found out by flexure formula

$$M = f_b \frac{I}{y}$$

M = Moment of resistance of the girder

I = Moment of inertia of girder

f_b = Maximum allowable bending stress

y = Extreme fibre distance
Moment of résistance of girder $= M = F \times D$
If f_b = Average bending stress
A = Flange area
$F = f_b \times A$
$M = f_b \times A \times D$

This expression, however does not take into account the bending moment contributed by the web and the error is on the safe side.

However, if the value for f_b is the maximum allowable stress in the extreme fibre stress, the error is on the unsafe side, since in the above formula f_b is the average stress at the centre of gravity of the flanges.

M_f = resisting moment of the flange $= A_f \times f_b \times D$ where A_f = area of cross-section of one flange (flange plate and angles)

M_w = resisting moment of the web plate = area of web plate × average stress × lever arm $= t_w \times \frac{D}{2} \times \frac{f_b}{2} \times \frac{2}{3} D = \frac{1}{6} f_b \times t_w D^2$

Total resisting moment

$$= M_f + M_w = A_f \times f_b \times D + \tfrac{1}{6} f_b \times t_w \times D^2 = f_b D \left(A_f + \frac{t_w D}{6} \right)$$

$$= f_b D \left(A_f + \frac{A_w}{6} \right) = f_b A D$$

where $$A = \left(A_f + \frac{A_w}{6} \right)$$

$\frac{A_w}{6}$ is sometimes called the web equivalent

For tension flange

$$A = \left(A_f + \frac{A_w}{8} \right)$$

2.12. Moment of Inertia Method

Welded girders : $M = f_b \times \frac{I}{y}$

$$I = 2 A_f \left(\frac{d_o}{2} \right)^2 + t_w . \frac{d_w^3}{12} + I_f$$

$$I = 2 A_f \frac{d_o^2}{4} + t_w \frac{d_w^3}{12} = A_f \frac{d_o^2}{2} + t_w \frac{d_w^3}{12}$$

$d_o = d_w$ = approximately $= d$

$$\therefore \; I = \frac{d^2}{2} \left(A_f + \frac{A_w}{6} \right)$$

Moment of resistance $= M = f_b \frac{I}{y} = f_b \frac{d^2}{2} \left(A_f + \frac{A_w}{6} \right) \times \frac{2}{d}$

$$= f_b d \left(A_f + \frac{A_w}{6} \right) = f_b d \times A$$

2·13. Specifications

2·13·1. Web Plate

Minimum thickness of web plate for unstiffened webs $=\frac{d}{85}$

For vertically stiffend web $=\frac{1}{180}$ of

the smallest clear panel dimensions $=\frac{d}{200}$

For webs with longitudinal stiffness at 2/5 a from compression flange$=\frac{d}{250}$

For webs with horizontal stiffner also at neutral axis$=\frac{d}{400}$

2·13·2. Flange Plates

Maximum outstand for compression flange$=16\,t$

Maximum outstand for tension flange$=20\,t$

Flange angles in riveted girder shall be at least 1/3 of the gross flange area.

In riveted girders, the thickness or width of flange plates may be varied as the bending moment changes from section to section.

However, since only a sigle plate is used in welded girders, it is not possible to curtail the flange plates as in a rivetted girder.

The change in flange plate size is accomplished by using various length plates of different thicknesses. These ptates are butt welded to form a continuous flange.

2·13·3. Flange Plate Curtailment

(i) Riveted girders:

Since the maximum bending moment occurs only at centre, the designed cross sectional area is required at centre only.

As the bending moment reduces from centre towards supports, the section designed for maximum bending moment can be reduced. This reduction can be met with by cutting out top plates at some points on the girder. This method of proportioning flange plates to achieve economy in the design is called the curtailment of flange plates.

Let M_c=bending moment at the centre, M_x=bending moment at any point K at a distance x from centre, A_{fc}=flange area required at centre, A_{fx}=flange area required at K at a distance x, w=load per unit length on the girder, L=effective span,

Since the bending moment is resisted by the flange plates, it can be assumed that flange area is proportional to bending moment,

$$\frac{A_c}{A_x}=\frac{M_c}{M_x}$$

$$\frac{A_c-A_x}{A_c}=\frac{M_c-M_x}{M_c} \quad \ldots(1)$$

$$M_c=\frac{wL^2}{8}$$

$$M_x=\frac{wL}{2}\left(\frac{L}{2}-x\right)-\frac{w}{2}\left(\frac{L}{2}-x\right)^2$$

$$M_x=\frac{wL^2}{8}-\frac{wx^2}{2}$$

Substituting in (1)

$$\frac{M_c-M_x}{M_c}=\frac{\frac{wL^2}{8}-\left(\frac{wL^2}{8}-\frac{wx^2}{2}\right)}{\frac{wL^2}{8}}$$

$$=\frac{4x^2}{L^2}=\frac{A_c-A_x}{A_c}$$

$$2x=L\sqrt{\frac{A_c-A_x}{A_c}}$$

Where $2x$ = length of flange plate = the distance at which the flange plate can be curtailed

$$l=L\sqrt{\frac{A_c-A_x}{A_c}}$$

$$l=L\sqrt{\frac{a}{A_c}}$$

Where $a=A_c-A_x$ = Area of the cover plate in question.

Generally three flange plates are used.

Let the areas of these plates be, a_1, a_2 and a_3 respectively.

Then, the lengths at which these can be curtailed is given by

$$l_1=L\sqrt{\frac{a}{A_c}}$$

$$l_2=L\sqrt{\frac{a_1+a_2}{A_c}}$$

$$l_3=L\sqrt{\frac{a_1+a_2+a_3}{A_c}}$$

The first flange plate should not, however be curtailed.

(*ii*) *Welded girders*: The change in flange plate size is accomplished by using various length plates of different thicknesses. These plates are butt welded to from a continuous flange.

If the difference in thickness of the two plates exceeds 25% of the thickness of the thinner plates or 3·8 mm, whichever is greater, the dimensions of the thicker part should be reduced at the butt joint to those of the smaller part, the slope being not steeper than 1 in 5.

If the reduction of thickness of thicker plate is impracticable or the joint is not subjected to dynamic load, the weld metal should be built up at the junction to dimensions at least 25% greater than the thinner part.

This results in a more uniform flow of stress.

2·14. Flange to Web Connection

The flanges of plate girders are connected either by rivets or welds to transmit the maximum horizontal shear force resulting from the bending moments on the girder.

(i) *Riveted girder*

Shear force $= S$; Depth of girder $= d$

Vertical shear per unit depth of girder $= \frac{S}{d}$ kg/cm

Least strength of rivet $= R$

Pitch of rivets $= p$

Resistance of the rivet per unit height $= \frac{R}{p}$ kg/cm

$$\text{Pitch of rivets} = \frac{S}{d} = \frac{R}{p}$$

$$p = \frac{Rd}{S}$$

or

$$p = \frac{RI}{SQ}$$

(ii) *Welded grider*

s = spacing between welds

W = weld value

S = Shear at a section

Q = Static moment about the centroidal axis of the portion of cross sectional area beyond the location at which the stress is desired.

I = moment of inertia

Shear stress at any point is $q_s = \frac{SQ}{IA_w}$

$$= \frac{SQ}{It_w \times 1} = \frac{SQ}{It_w}$$

or spacing s of the weld $= \frac{WI}{SQ}$

Size of the weld $= s_w$ cm

Assuming that the weld is continuous and used on both sides

Strength of weld per centimetere length $= 2\, s_w \times 0{\cdot}7 \times 1025$ $= 1450\ s_w$

Where 1025 kg/cm² is the allowable shear stress in the weld

$1450\ s_w$ = Shear force per centimeter height from which size of the weld can be calculated.

2·15. Minimum Size for Fillet Welds

Thickness of thicker plate	*Specified minimum size of the weld*
Upto and including 9·5 mm	3 mm
Over 9·5 mm and upto 19 mm	5 mm
Over 19 mm and to 32 mm	6 mm
Over 32 mm and 50 mm	9·5 mm

For plates thicker than 50 mm special precautions are to be taken.

2·16. Maximum Size of Fillet Welds

The maximum effective size of fillet weld is limited by the strength of web plate in shear

$$2\,s w_{max} \times 0{\cdot}7 \times 1025 = tw \times 945$$

$$s_{max} = 0{\cdot}652\ tw = \frac{2}{3} \text{ the web thickness approximately}$$

2·17. Stiffners

(*i*) Intermediate stiffners.

Intermediate stiffners are required to prevent the web plate from buckling under a complex and variable stress situation resulting from combined shear and bending moment.

These stiffners have a secondary function not generally recognised in that if fitted against the flanges at top and bottom, they maintain the perfect 90° angle between flanges and the web.

These stiffners are provided when $\frac{d}{t} > 85$ or thickness of web $t_w < \frac{d}{85}$ spacing of these stiffners $C \nless \frac{d}{5} \ngtr 1{\cdot}5\,d$

Moment of intertia of the stiffner $I_s \nless \dfrac{1{\cdot}5\, d^3\, t^3}{c^2}$

The width of the stffner should not exceed 16 times thickness of the stiffner.

d = distance between the flange angles for riveted girders.

d = clear distance between flange plates ignoring fillets for welded girder.

= clear distance between the horizontal stiffner and the tension flange ignoring fillets when horizontal stiffner is provided.

I = The moment of inertia of a pair of stiffners about the centre of the web or of a single stiffner about the face of the web.

t = The minimum required thickness of the web.

c = The maximum permitted clear distance between vertical stiffner for thickness t.

Intermediate vertical stiffners may be in pairs arranged one on each side of the web or single. When single stiffners are used, they should be preferably placed alternately on opposite sides of the web. The stiffners shall extend from flange to flange and may be joggled over the vertical legs of the flange angles.

2·18. Web Panel Dimensions

In no case shall the greater unsupported clear dimensions of a web panel exceed 270 t_w nor the lesser unsupported clear dimensions of the same panel exceed 180 t_w. where t_w=thickness of the web plate.

2·19. Outstand of the Stiffners

For sections other than flats, the outstand of all stiffners from the web shall not be more than 16 t.

For flats outstand shall not be greater than 12 t.

Where t=thickness of flat or section.

2·20. Bearing Siffner

The function of the bearing stiffner is to transmit concentration of load so as to avoid local bending failure of the flange and local crippling or buckling of the web.

Selection of stiffnerss is usually made on basis of local permissible contact bearing pressures of 1890 kg/cm^2 at the points of bearing contact between the outstanding parts of the bearing stiffners and the flanges.

Welds or rivets shall be used to transfer the total load from the bearing stiffners to the web.

The bearing stiffner together with the web plate shall be designed as a column with an equivalent reduced slenderness ratio.

Effective length=0.7 actual length.

The area of the section which resists compression is the pair of stiffners together with a length of web on each side of the centre line of the stiffners where available = 20 times the web thickness.

The bearing stiffners at the support points should project as nearly as possible to the outer edges of the flanges.

In the case of riveted bearing stiffners filler plates shall be used.

2·21. Longitudinal Stiffners

In addition to the vartical intermediate stiffners, bearing stiffners shall be provided if $t<d/200$, one horizontal stiffner shall be placed on the web at a distance from the compression.

Flange = 2/5 of the distance from the compression flange to the neutral axis when the thickness of the web is

$$<\frac{t_w}{250}$$

These stiffners shall be designed so that $I \nless 4c_1 t^3$

when $d/t > 250$, a second horizontal stiffner shall be placed at neutral axis.

In this case $I \nless dt^3$.

Horizontal web stiffners shall extend between vertical stiffners but need not be continuous over them.

Horizontal stiffners may be in pairs arranged on each side of the web or single.

2·22. Stiffner Connections

The outstanding plates may be directly welded to the web by fillet or butt welding. Minimum size of welds depends upon the thickness of web or stiffner or as required for shear force.

$$\frac{t_w^2}{2h} \text{ tonnes per cm should usbe ed}$$

where h = maximum outstand of the stiffner.

If the intermittant welds are used, the distance between the effective lengths of any two weld, even if staggered on opposite sides of the stiffners should not exceed 16 times the thickness of the stiffner nor 30 cm.

If the welding is done on one side of stiffner or on two sides but staggered the effective length of each weld should not be less than four times the thickness of the stiffner.

Where, rivets are used, the minimum pitch $\nless 2.5d$

where d = diameter of the hole and the maximum pitch $\ngtr 32t$ or 300 mm whichever is less where

t = the thickness of the thinner outside plate.

3. Welded Plate Girder for Multi-storey Departmental Store 20 m span

3·1. Loads (Dead loads)

Roof slab (10 mm)	= 240 kg/m²	Floor slab (10 cm)	= 240 kg/m²
Roof finish	= 110 ,,	Floor finish	= 60 ,,
	350 kg/m²		300 kg/m²

3·2. Live Load

On roof = 150 kg/m²

On floors (office) = 250 kg/m²

3·3. Column Load

Area supported by each column resting on plate girder

$= 5 \times 2{\cdot}5 = 12{\cdot}5$ sq. m

Loads on the column,

Due to dead loads: Roof = 350 kg/m²

Floors (6 Nos) = $300 \times 6 = 1800$ kg/m²

Total DL = 2150 kg/m²

Due to live loads: Roof live ioad = 150 kg/m²

Floor live loads

(column carries 6 floors) = $0{\cdot}6 \times 250 \times 6 = 900$ kg/m²

(For column carrying 5 floors and above

Reduction factor on LL is 0·4)

Total live load = 1050 kg/m²

Total load due to DL + LL = 2150 + 1050 = 3200 kg/m²

= 3·2 t/m²

Load on the column = $3{\cdot}2 \times 12{\cdot}5 = 40$ t

Self wt of secondary beam (5 m long) at 75 kg/m = $5 \times 75 \times 7$

= 2625 kg = 2.625 t

Main beam at 200 kg/m (2·5 m long) = $2{\cdot}5 \times 150 \times 7 = 2625$ kg

= 2·625 t

Total load on the column = 40 + 2·625 + 2·625 = 45·25 t

Self weight of the column at 10% = 4·525 t

49·775 t

Say = 50 t

3·4. Loading on the plate grinder

The girder carries three concentrated loads through the columns of the upper floors and seif weight of the girder of ½ t per metre on the entire length.

3·5. Maximum Bending Moment

$$\text{Reaction} = \frac{\text{total load}}{2} = \frac{150 + 20 \times \frac{1}{2}}{2} = \frac{160}{2} = 80 \text{ t}$$

Bending moment under load (2) that is at entire of the span
$=80\times 10-50\times 5-\frac{1}{2}\times 10\times \frac{10}{2}=800-250-25$
$M_e=525$ mt.

3·6. **Shear force**

Maximum shear force $=80$ t

3·7. **Web plate**

Approximate depth of girder (heavy loading) $d=\frac{1}{8}$ of span $=\frac{1}{8}$ of $2000=250$ cm

Economical depth $=d=5\sqrt[3]{\frac{M}{f_b}}=5\sqrt[3]{\frac{52500{,}000}{1500}}$

$=163{\cdot}75$ cm

Depth of girder required for shear assuming 6 mm (minimum thickness) thick web $\frac{80{,}000}{945\times{\cdot}6}=141$ cm

Use 160 cm $\times$ 0·6 cm web plate

3·8. **Flange plates**

Approximate flange area required $=A_f=\frac{M}{F_b\, d}-\frac{A_w}{6}$

$=\frac{52500{,}000}{1500\times 160}-\frac{1}{6}\times 160\times 0{\cdot}6=219-16=203\ \text{cm}^2$

Flange width $=\frac{1}{40}$ or $\frac{1}{45}$ span $=\frac{1}{40}\times 2000=50$ cm

Approximate thickness of plate required $=\frac{203}{50}=4$ cm

Flange width $\not> 32\ t=32\times 4=128$ cm
Hence use 50 cm $\times$ 4 cm flange plates.

3·9. **Self Weight Check**

Weight of web plate/m length $=1{\cdot}60\times 1\times\frac{0{\cdot}6}{100}\times 7850$
$=75{\cdot}2$ kg/m

Weight of flange plates/m length $=2\times\frac{50}{100}\times\frac{4}{100}\times 1\times 7850$
$=315$ kg/m

Weight of web and flanges plates/m length $=390{\cdot}2$ kg/m
Add 30% of the web plate weight for welded Girders (additional weight due to welding etc.) $=75{\cdot}2\times\frac{30}{100}=22{\cdot}6$ kg/m

Self weight per metre length of girder $=412{\cdot}8$ kg/m
<500 kg/m assumed

3·10. **Connection between Flange and Web**

Maximum shear at the junction of web and Flange $q=\frac{F a y}{2}$

$F=80000$ kg, $a=50\times 4=200\ \text{cm}^2$, $y=82$ cm.

Since

$$I = \frac{50}{12}(168^3 - 160^3) + \frac{0.6 \times 160^3}{12}$$

$$= 2700{,}000 + 144800 = 284{,}4{,}800 \text{ cm}^4 = 2{,}8448 \times 10^6 \text{ cm}^4$$

$$q = \frac{80{,}000 \times 50 \times 4 \times 82}{2844800} = 462 \text{ kg/cm}$$

Size of weld assuming continuous weld on either side, required $= 2 \times \cdot 7 \times s \times 1025 = 1450s = 462$

$$s = \frac{462}{1450} = 0{\cdot}319 \text{ cm}$$

Use 5 mm fillet weld continuous on either side.

3·11. Check for maximum stresses

$$\text{Moment of inertia} = I = \frac{0{\cdot}6 \times 160^3}{12} + 2 \times 4 \times 50\,(82)^2 + 2 \times \frac{1}{12} \times 50 \times 4^3$$

$$= 20{,}4{,}800 + 2689{,}600 + 533{\cdot}33 = 289{,}49{,}33{\cdot}33 \text{ cm}^4$$

$$= 2{,}89493 \times 10^6 \text{ cm}^4$$

3·12. Bending stress

Bending stress

$$f_b = \frac{52500000 \times 82}{2894933{\cdot}33} = 1485 \text{ kg/cm}^2 < 1575 \text{ kg/cm}^2$$

3·13. Shear stress

$$\text{Shear stress} = \frac{80{,}000}{160 \times 6} = 833{\cdot}3 \text{ kg/cm}^2 < 945 \text{ kg/cm}^2$$

3·14. Stiffners

$$\frac{d}{t} = \frac{160}{0{\cdot}6} = 266{\cdot}6 > 250$$

Hence two longitudinal stiffners are needed in addition to the vertical intermediate stiffners.

3·15. Intermediate Stiffners

$$\frac{d'}{t} = \frac{80}{0{\cdot}6} = 133{\cdot}3 > 85$$

Spacing of intermediate stiffners $\frac{d'}{3}$ or $1{\cdot}5\,d' = 27$ cm to 120 cm

Use 100 cm c/c.

3·16. Check for Panel Dimensions

Greatest unsupported panel dimension of the web = 100 cm $< 270\,t_w = 270 \times 0{\cdot}6 = 162$ cm.

Lesser unsupported clear panel dimension of the web = 80 cm $< 108\ t_w = 180 \times 0{\cdot}6 = 108$ cm.

The stiffners are designed to have minimum moment of inertia

$$=I=\frac{1\cdot5\ d'^3\ t_w^3}{C^2}=\frac{1\cdot5\times80^3\times0\cdot6^3}{(100)^2}=16\cdot58\ \text{cm}^4$$

Use 0·6 cm plate.
Maximum width of stiffner plate$=16\,t=16\times{\cdot}06=9{\cdot}6$ cm
Use 8 cm × 0·6 cm plate
I provided$=\frac{1}{3}\times0{\cdot}6\times8^3=102{\cdot}4\ \text{cm}^4 > 16{\cdot}58\ \text{cm}^4$.

3·17. Connection of vertical Stiffner to Web

Shear on welds connecting stiffner to web

$$=\frac{t_w^2}{2h}\ \text{tonnes/cm}=\frac{(0\cdot6)^2}{2\times8}={\cdot}0225\ \text{t/cm}=22\cdot5\ \text{kg/cm}$$

$$\text{Size of the weld}=\frac{22\ 5}{0\cdot7\times1025}={\cdot}0314\ \text{cm}$$

minimum size of the weld = 3 mm.
Effective length of weld $\nless$ 10 $t=10\times0{\cdot}6=6$ cm.
Use 8 cm in length and 3 mm size fillet weld.

3.18. Bearing Stiffner

Maximum shear force$=25+80=105\ t$

$$\text{Bearing area required}=\frac{105000}{1890}=56\ \text{cm}^2$$

The size of the bearing plate required is very small and a suitable column cap shall be provided.

Bearing stiffners should be extended as nearly as practicable to the outer edges of the flanges

Use 24 cm wide plates

$$\frac{b}{t}\ngtr12$$

$$t=\frac{24}{12}=2\ \text{cm}$$

Use 24 cm × 2 cm plate
Bearing area of the stiffners alone$=24\times2\times2=96\ \text{cm}^2>56\ \text{cm}^2$

The length of web plate which acts along with the stiffner plates in bearing the reaction $=20\,t=20\times0{\cdot}6=12$ cm

On either side . Area of web$=2\times12\times0{\cdot}6=14{\cdot}4\ \text{cm}^2$

Total area $=96+14{\cdot}4$

$$\text{I for the section}=110\cdot4\ \text{cm}^2=\frac{2\times(48\cdot6)^3}{12}+\frac{2\times11\times0\cdot6^3}{12}$$

$$=19200+0\cdot3=19200\cdot3\ \text{cm}^4$$

$$\text{radius of gyration}=\sqrt{\frac{I}{A}}=\sqrt{\frac{19200\cdot3}{110\cdot4}}=13\cdot2\ \text{cm}$$

Effective length$=0{\cdot}7\times160=112$ cm

$$\frac{l}{r}=\frac{112}{13\cdot2}=8\cdot5$$

Allowable stress in compression$=1246{\cdot}6\ \text{kg/cm}^2$

Area required to resist the reaction$=\dfrac{105,000}{1246{\cdot}6}=84$ cm^2

$< 110{\cdot}4$ cm^2 provided

3·19. Weld connection

Length available for welding$=2(160-4)=312$ cm

Required strength of weld$=\dfrac{105000}{312}=336$ kg/cm^2

Size of weld $=\dfrac{336}{0{\cdot}7\times1025}={\cdot}33$ cm

Use 5 mm weld

Length of weld $\nless 10\ t=10\times0{\cdot}6=6$ cm

Use 10 cm length and 5 mm size alternate on either side of the stiffner plate and on both sides of the web.

3·20. Stiffners under concentrated loads

Load transmitted by column$=50\ t$

Bearing area required $=\dfrac{50,000}{1890}=25{\cdot}5$ cm^2

Using 10 cm$\times$2 cm with 2$\times$2 cm chamfer at the ends.
Area availabie for three plates$=8\times2\times3=48$ cm^2
Area of web$=24\times0{\cdot}6=14\ 4$ cm^2
Total area $=62{\cdot}4$ cm^2
Moment of inertia of the arrangement neglects web contribution

$$I=\frac{3\times2\times(10{\cdot}3)^3}{3}=2183{\cdot}6\ \text{cm}^4$$

$$r=\sqrt{\frac{I}{A}}=\sqrt{\frac{2183{\cdot}6}{62{\cdot}4}}=5{\cdot}92\ \text{cm},\ l=0{\cdot}7\times160=112\ \text{cm}$$

$$\frac{l}{r}=\frac{112}{5{\cdot}92}=18\ 9$$

Allowable stress $f_c=1239{\cdot}77$ kg/cm^2

Area required $=\dfrac{50,000}{1239\ 77}=40{\cdot}3$ cm^2 $< 62{\cdot}4$ cm^2 provided

3·21. Connections

Strength of weld required$=\dfrac{50,000}{2\times(160-4)}=106{\cdot}5$ kg/cm^2

Size of weld required$=\dfrac{106{\cdot}5}{0{\cdot}7\times1025}=0{\cdot}149$ cm

Use minimum size of the weld$=0{\cdot}3$ mm.
Length of the weld $10\times t=10\times{\cdot}6=6$ cm
Use 10 cm alternately staggered on either side

3·22. Longitudinal Stiffners

$$\frac{d_w}{t}=\frac{160}{0{\cdot}6}=266\ 6>250$$

Two horizontal stiffners shall be provided. The first stiffner at a

WELDED PLATE GIRDER FOR A MULTI STORY DEPARTMENTAL STORE - 20M SPAN

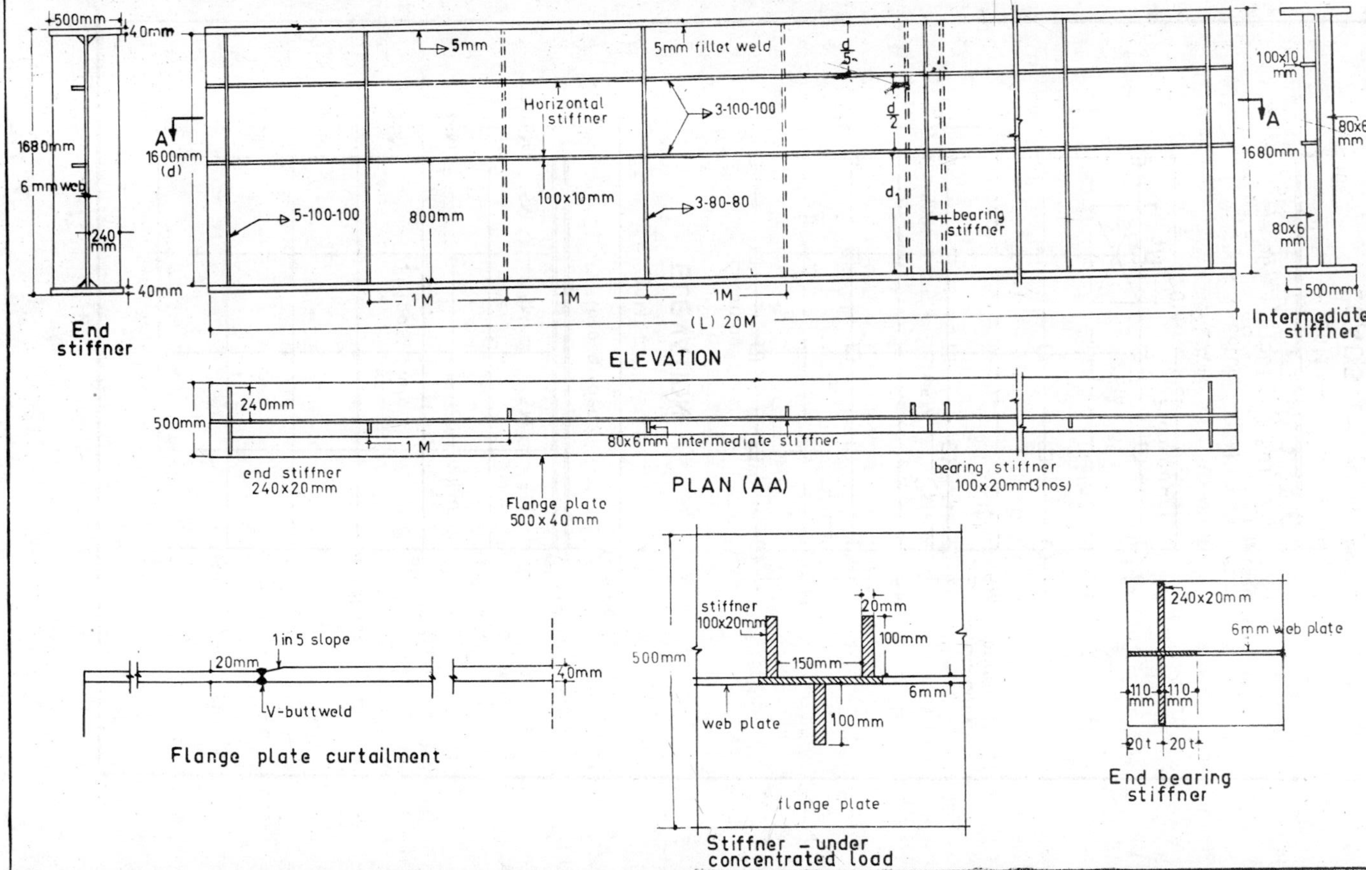

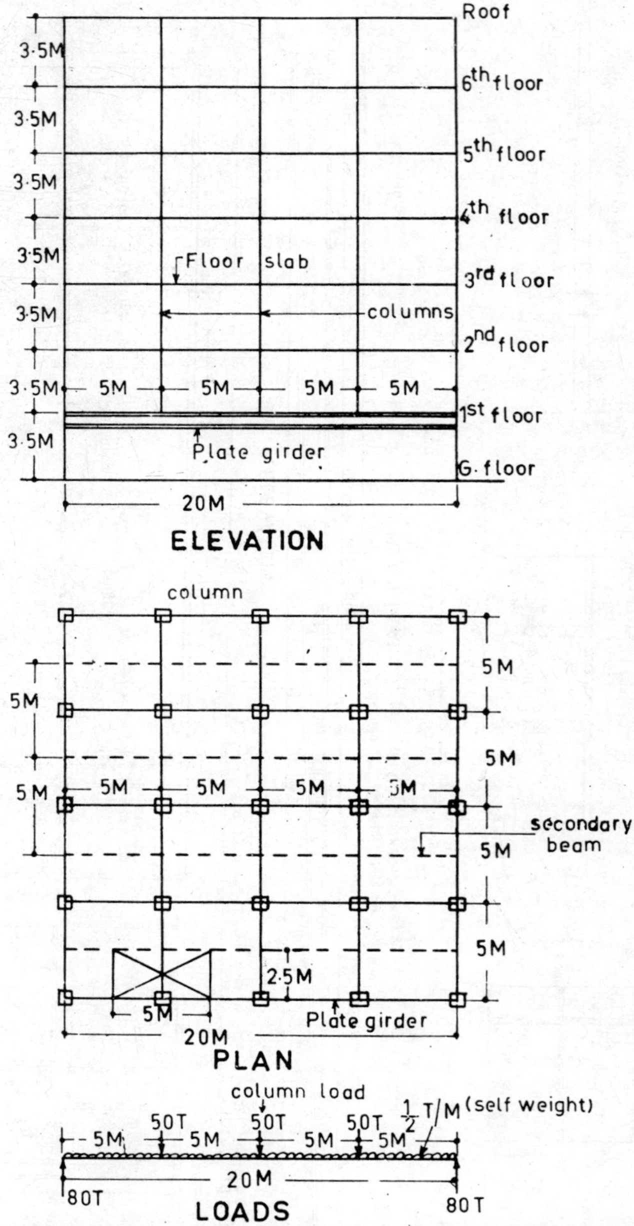
Roof
6th floor
5th floor
4th floor
3rd floor
2nd floor
1st floor
G. floor
3.5M
Floor slab
columns
5M
Plate girder
20M
ELEVATION
column
secondary beam
2.5M
PLAN
column load
50T
1/2 T/M (self weight)
80T
LOADS

depth of $d/5$ from compression flange and the second at $d/2$ (Neutral axis).

Stiffener at $\frac{d}{5}$ level

Moment of inertia required

$$=I=4\ C_1\ t_w^3$$
$$=4\times 100\times(0{\cdot}6)^3$$
$$=86{\cdot}4\ \text{cm}^4$$

Using 10 cm wide 1 cm thick plates

Moment of inertia

$$=\tfrac{1}{3}\times 1\times 10^3=333{\cdot}3\ \text{cm}^4$$
$$> 86{\cdot}4\ \text{cm}^4\ \text{required}$$

Stiffener at N. A

I required $=d'\ t^3$

$$=128\times(0{\cdot}6)^3$$
$$=27{\cdot}6\ \text{cm}^4$$

Use 6 cm wide 0·6 cm plate.

3.23. Connections

Shear force in tonnes per cm run of weld

$$=\frac{t^2}{2h}\ \text{t/cm}$$
$$=\frac{0{\cdot}6^2}{2\times 6}\times 1000=30\ \text{kg/cm}$$

Fillet weld size required

$$=\frac{30}{0{\cdot}7\times 1025}$$
$$=\cdot 0416\ \text{cm}$$

Weld size required at $d/5$

$$\text{Shear/cm length}=\frac{1^2}{2\times 10}=\frac{1}{20}\ \text{t/cm}$$
$$=50\ kg/\text{cm}$$

$$\text{Size of weld required}=\frac{50}{0{\cdot}7\times 1025}$$
$$=\cdot 069\ \text{cm}$$

Use minimum size of weld of 3 mm—

Length of weld $\nless 10\times t$

Use 10 cm length welds staggered on either side.

PART II

DESIGN OF HIGHWAY BRIDGES

1. Highway Bridges—General

1·1. Types

(*a*) Solid web girder bridges—(*i*) Rolled steel girder bridges (*ii*) plate girder bridges.

(*b*) Open web girdes bridges—Trussed girder bridges.

Rolled steel girder bridges are used upto 10 m spans. For 10 to 40 m spans plate girder bridges are used. Trussed girder bridges are used for spans ranging from 30 m and above. These bridges may be through or deck type.

1·2. Loadings

Steel bridges are designed using standard specifications and code of practice for road bridges-Section V and loading standards as specified in section II of the Indian Road Congress. Various loading on bridges are IRC Class AA, IRC Class A and IRC Class B.

1·3. Dead Loads

Self weight of girders

For Plate girders of 10 m to 30 m span, $w=20\ L+100$ kg/m

For trussed girder bridges 30 m to 90 m, $w=15\ L+550$ kg/m

1·4. Impact Factors

Class A and B Loading

Impact factor for spans upto 3 m=54·5%

Impact factor for span from 3 m to 45 m=$\dfrac{9}{13{\cdot}5+L}$

For Class AA loading

(*i*) Less than 9 m spans;

For tracked vehicles=25% for spans upto 5 m, linearly reducing to 10% for spans of 9m

For wheeled vehicles=25%

(*ii*) 9 m spans and over;

For tracked vehicles—10% for all spans

For wheeled vehicles—25% for spans upto 23 m and in accordance with the curve for spans in excess of 23 m.

1·5. Lateral Wind Force

The lateral wind force of 300 kg/m against any exposed moving live load shall be considered as acting at 1·5 m above the roadway.

The bridges shall not be considered to be carrying any live load when the wind velocity at deck level exceeds 130 km per hour.

1·6. Longitudinal Forces

The braking effect on a simply supported span or a continuous unit of spans or any other type of bridges for a single or two lane bridge, 20% of the first train load plus 10% of the load of the succeeding trains or part thereof, the train loads in one lane only being considered.

Where the entire first train is not on the full span, the braking force will be taken as equal to 20% of the loads actually on the span.

For bridges having more than two lanes the loads calculated as per two lanes plus 5% of the loads on lanes in excess of two.

The force due to braking effect shall be assumed to act along a line parallel to the roadway and 1·2 m above it, While transferring the force to the bearings, the change in the vertical reaction at the bearings should be taken into account.

The longitudinal force at any free bearing shall be limited to the sum of DL and LL reactions at the bearing multiplied by the appropriate co-efficient of friction, The co-efficient of friction at the bearings shall be assumed as follows.

For roller bearings	0·03
For sliding bearings of hard copper alloy	0·15
For sliding bearings of steel on cast iron or steel on steel	0·25
For sliding bearings of steel on ferro asbestos.	0·20

1·7. Permissible Stresses

Permissible stresses in steel

(*i*) Axial tension	1410 kg/cm²
(*ii*) in bending	
Plates, flats, tubes, rounds square and similar sections	1570 kg/cm²
Rolled beams, channels, angles, tees and for plate girders ($d/t<85$)	1500 kg/cm²
For plate girders ($d/t>85$)	1410 kg/cm²
(*iii*) in shear	
Maximum shear streess	1020 kg/cm²
Average shear stress, webs of plate girders, rolled beams, channels, angles and tees	850 kg/cm²

(*iv*) in compression

l/r	P_{ac} kg/cm²
0	1400
20	1360
40	1300
60	1180
80	1010
100	805
120	630
140	494
160	390

1·8. Bolts

In tension in Bolts

diameter exceeding 38 mm	1410 kg/cm²
diameter exceeding 28 mm but not 38 mm	1260 kg/cm²
diameter exceeding 22 mm but not 28 mm	1100 kg/cm²
diameter less than 22 mm	940 kg/cm²

1·9. Rivets

In tension in rivets

	in shear kg/cm²	in bearing kg/cm²
(*i*) Power driven shop rivets and turned and fitted bolts	1020	2360
(*ii*) power driven field rivets	940	2200
(*iii*) hand driven rivets	870	1890
(*iv*) black bolts	790	1580
(*v*) on flat surfaces	—	1890

1·10. Pins

	kg/cm²
In shear	1020
bearing	2120
Knuckle pins	1180

1·11. Bearings

Cylindrical rollers on curved surfaces

single and double rollers $\dfrac{0.8}{\left(\dfrac{1}{D_1}-\dfrac{1}{D_2}\right)}$ kg/mm of length

Three or more rollers $\dfrac{0.5}{\dfrac{1}{D_1}-\dfrac{1}{D_2}}$ kg/mm of length

Cylindrical rollers—cast steel. Single and double rollers

$$\frac{1.1}{\left(\dfrac{1}{D_1}-\dfrac{1}{D_2}\right)} \text{ kg/mm of length}$$

Three or more rollers $\dfrac{0.7}{\left(\dfrac{1}{D_1}-\dfrac{1}{D_2}\right)}$ kg/mm of length

where D_1 and D_2 are the diameters of the convex and concave contact surfaces respectively.

Cylindrical rollers on flat surfaces	Mild steel kg/mm length	Cast steel kg/mm length
single and double rollers	$0.8\ D_3$	D_3
Three or more rollers	$0.5\ D_3$	$0.7\ D_3$

where D_3=diameter of the roller.

1·12. Design

The effective span shall be

1·12·1. Effective Spans

(*i*) for main girders	—centre to centre of bearing plates on rocker pins.
(*ii*) for cross member	—distance between the centres of the main girders or trusses.
(*iii*) for road bearers or stringers	—distance between the centres of the cross members.
(*iv*) for deck slab	—the distance between the centres of the road bearers.

1·12·2. Effective Depth

The effective depth of plate or truss girder should be taken as the distance between the centres of gravity of the upper and lower flanges or chords.

1·12·3. Minimum Depth

The minimum depth preferable shall not be less than

(*i*) Trusses $=\frac{1}{10}$ of effective span

(*ii*) For rolled steel joists and plate girders$=\frac{1}{25}$ of effective span.

1·12·4. Spacing of Main Griders

The distance of centres between the main girders should be sufficient to resist overturning. The distance shall not be less than $\frac{\text{Span}}{20}$.

The depth between the gravity axes of the top and bottom chord of open web girders shall not be greater than three times the distance between the centres of main girders.

1·12·5. Minimum Sections

Minimum thickness for plates and rolled sections when used for main members.

(*i*) when accessible for painting on both sides=8 mm

(*ii*) when accessible for painting on one side=10 mm

Minimum size of the angle section $75\times50\times8$ or $75\times50\times10$ mm shall be used for main girders or trusses in riveted construction.

No angle less than $65\times45\times8$ or $65\times45\times10$ mm or flats less than 50 mm in width should be used in any part of the structure except hand railing.

1·12·6. Bearings

All bearings shall be designed to permit inspection and maintenance.

For spans greater than 15 m on rigid piers or abutments, bearings which will permit angular deflection of the girder ends shall be provided at one end, there shall be a rocker, roller or other effective type of expansion bearings.

1·12·7. Struts

Effective length of compression member $=0 \cdot 7\ L$ where $L=$ length of strut

For battened struts effective length be increased by 10%.

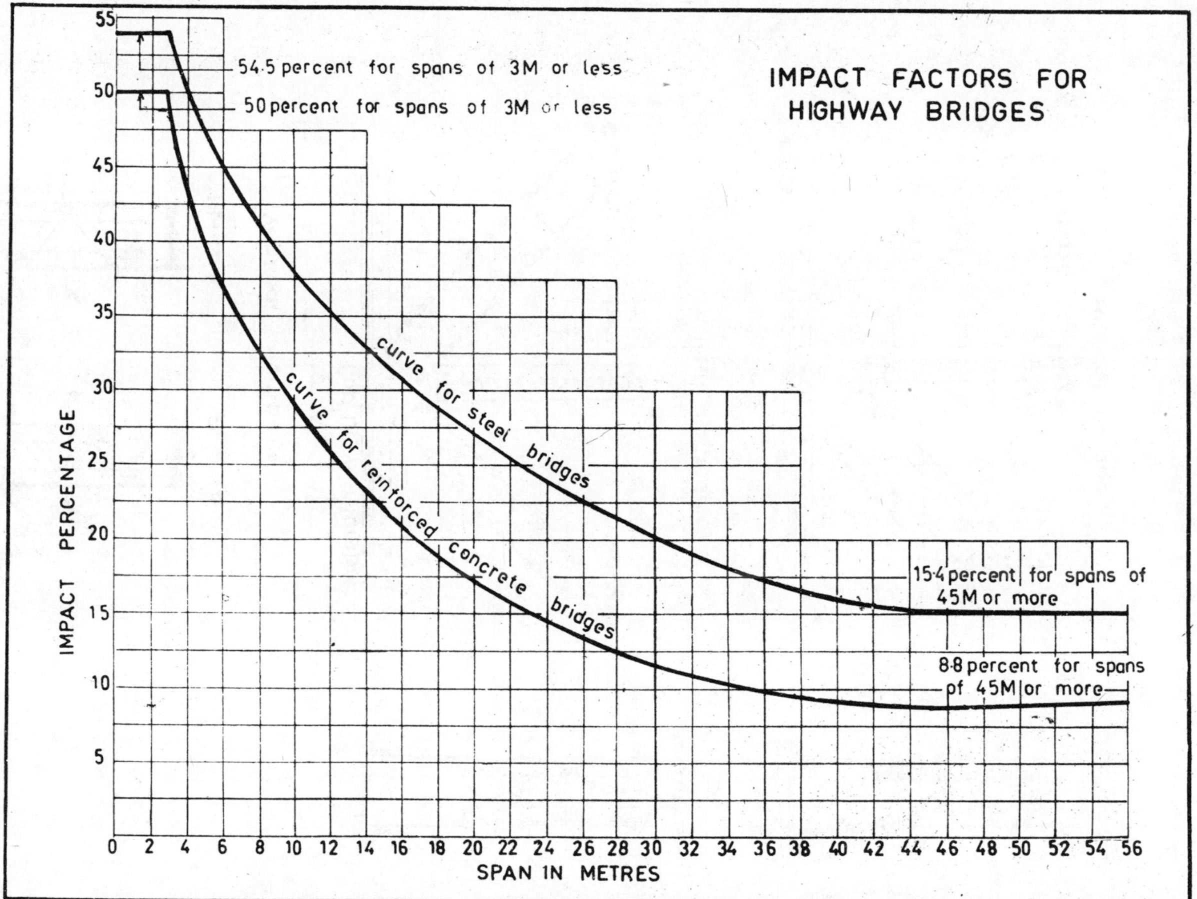
IMPACT FACTORS FOR
HIGHWAY BRIDGES
54·5 percent for spans of 3M or less
50 percent for spans of 3M or less
curve for steel bridges
curve for reinforced concrete bridges
15·4 percent for spans of
45M or more
8·8 percent for spans
of 45M or more
IMPACT PERCENTAGE
55
50
45
40
35
30
25
20
15
10
5
0 2 4 6 8 10 12 14 16 18 20 22 24 26 28 30 32 34 36 38 40 42 44 46 48 50 52 54 56
SPAN IN METRES

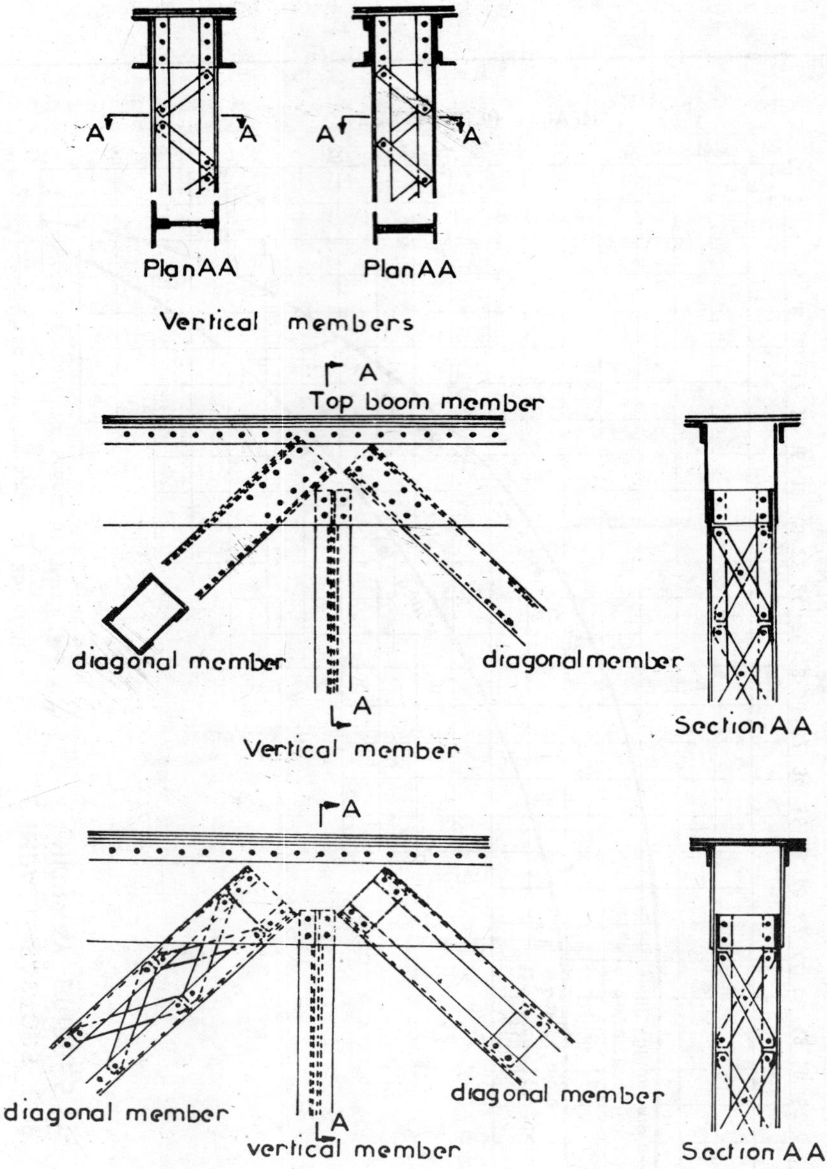

TOP BOOM CONNECTIONS

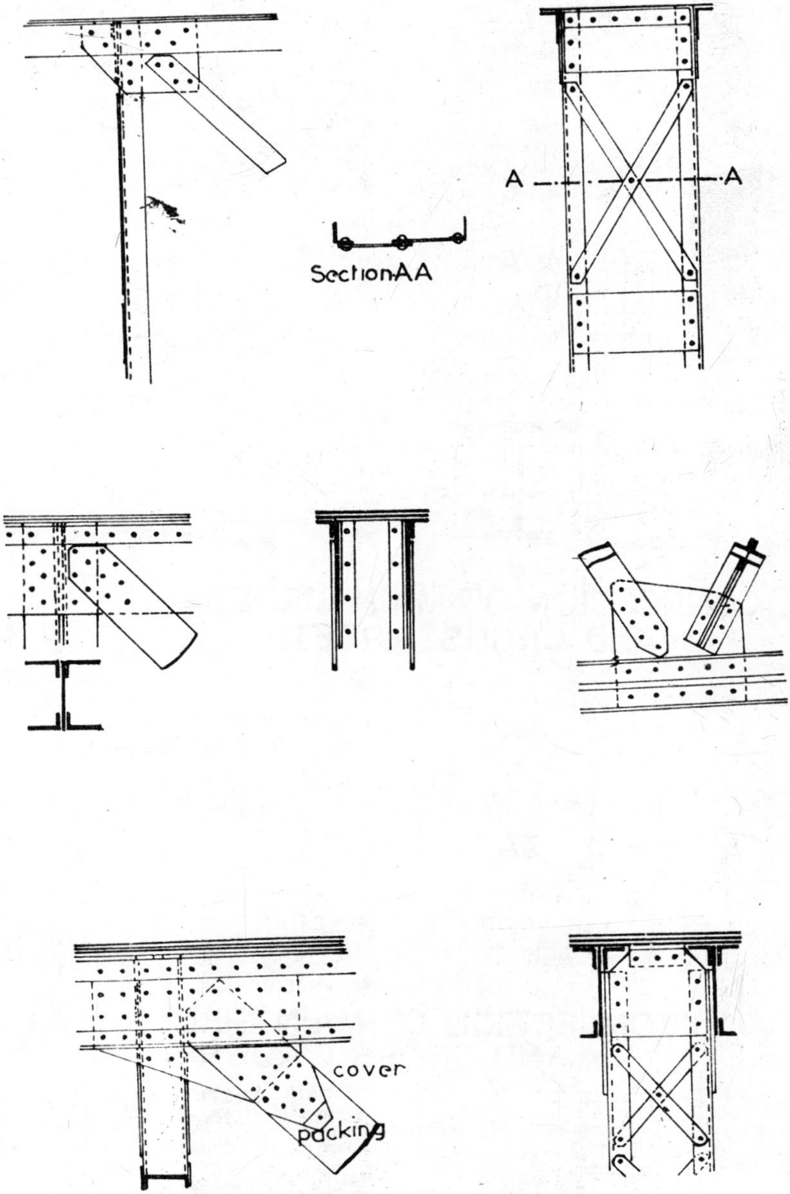

CONNECTION OF
WEB BRACEING WITH BOOMS

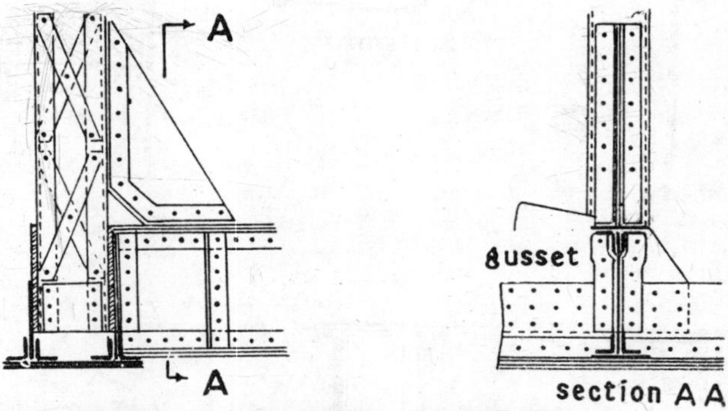

CONNECTION OF MAIN GIRDER
AND CROSS GIRDER

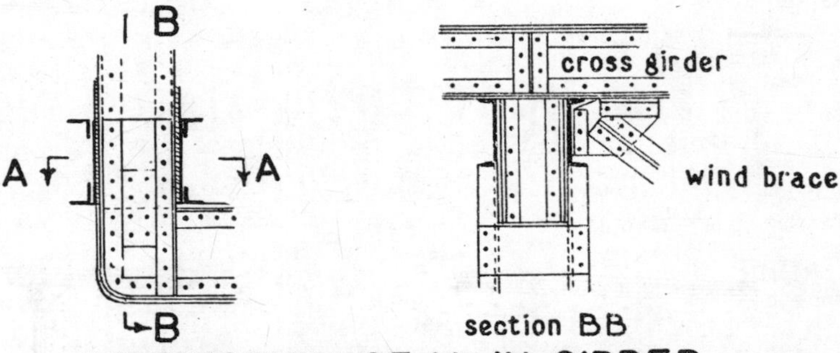

CONNECTION OF MAIN GIRDER
AND CROSS GIRDER

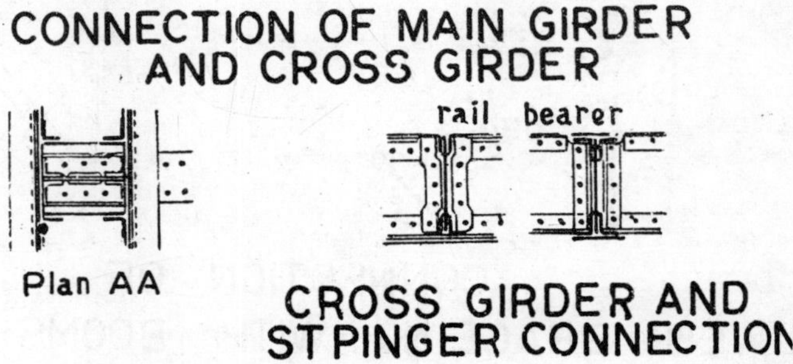

CROSS GIRDER AND
STPINGER CONNECTION

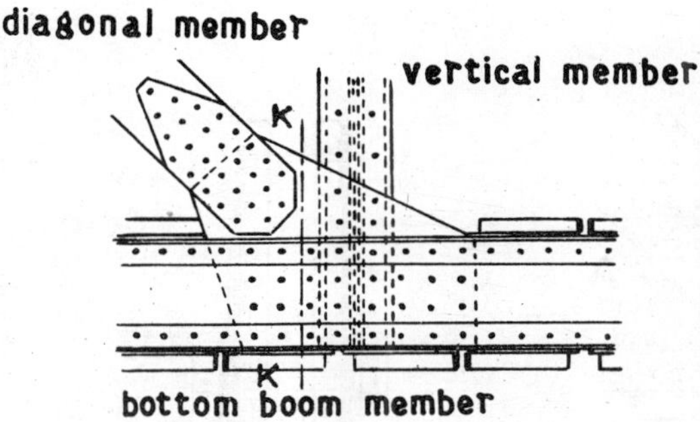

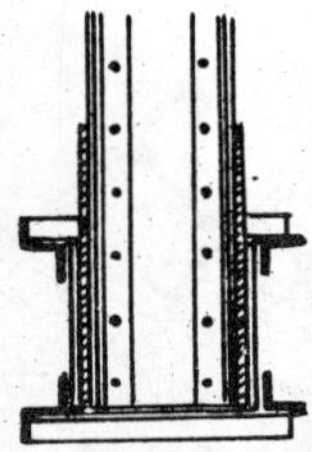

Section KK

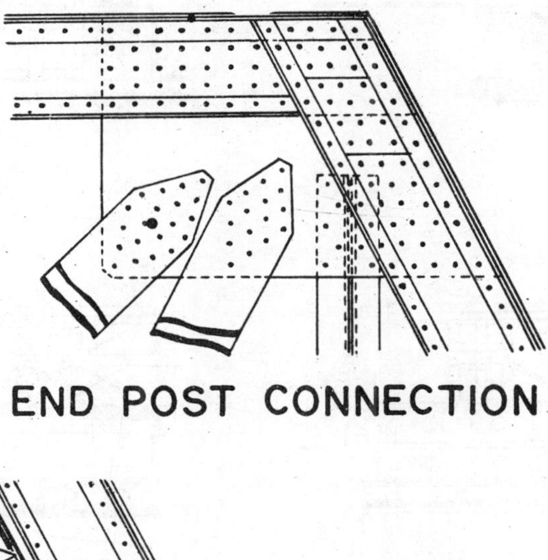

END POST CONNECTION

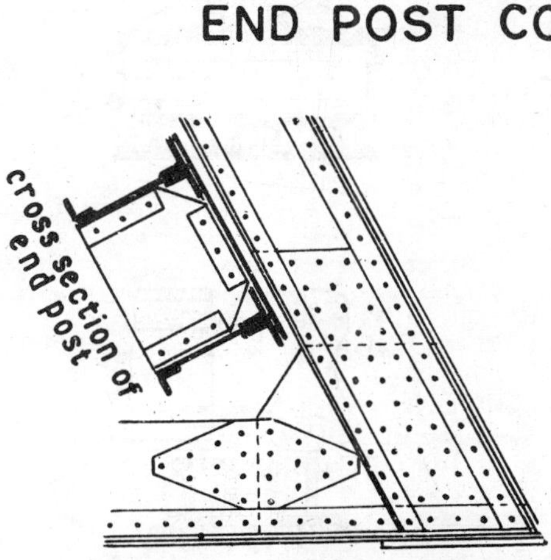

CROSS GIRDER AND
END POST

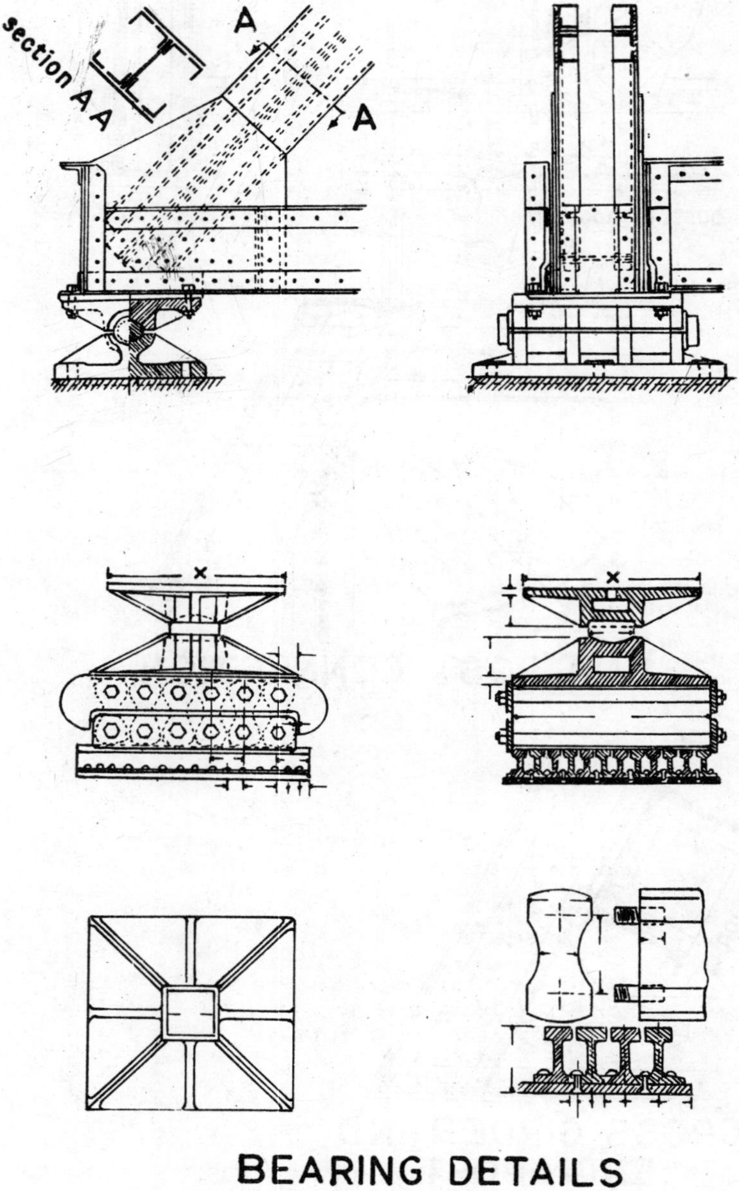

BEARING DETAILS

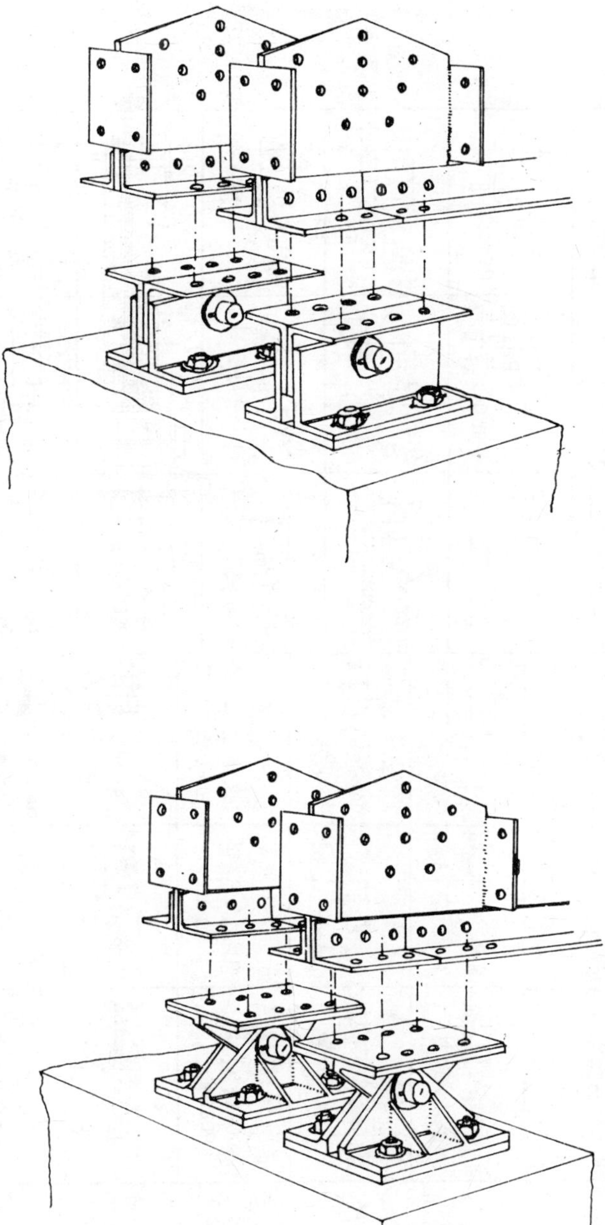

END BEARING

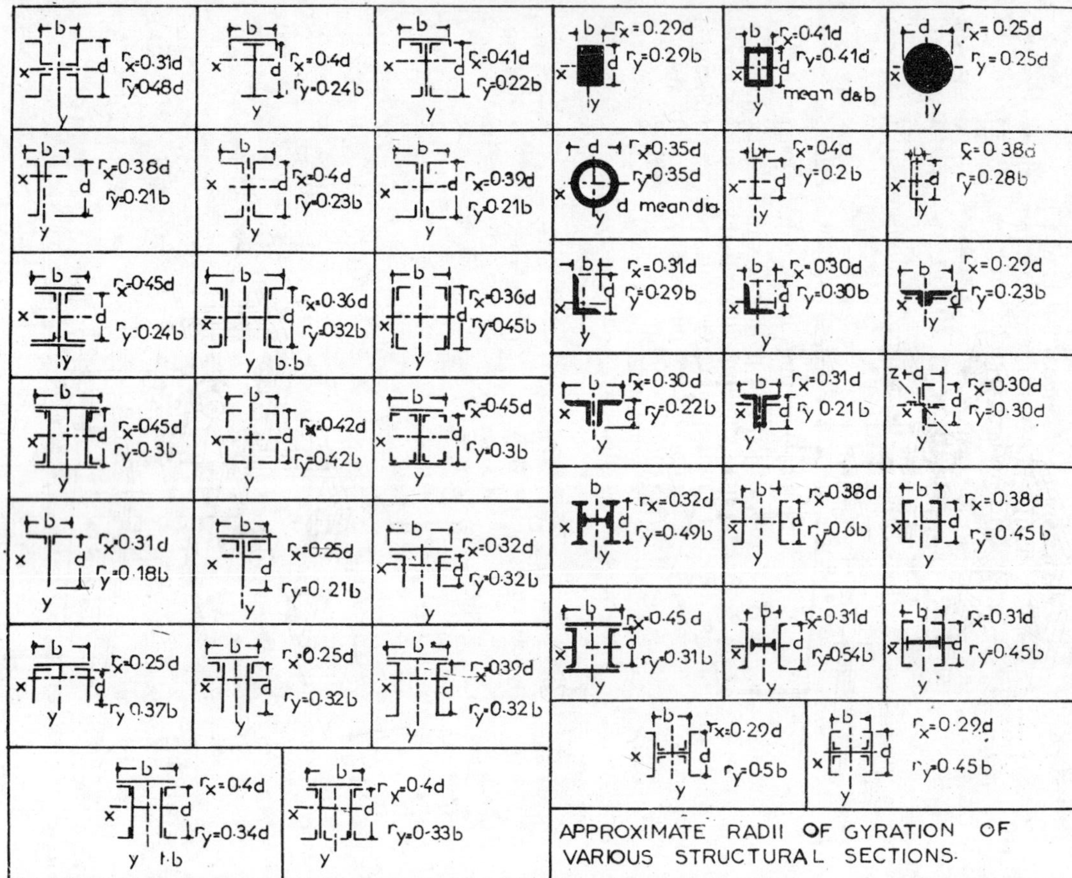

APPROXIMATE RADII OF GYRATION OF VARIOUS STRUCTURAL SECTIONS.

2. Design of Plate Girder Two-Lane Highway Bridge—Span 20 m

2·1. Data

Type—Semi through; Roadway—Two Lane (6·8 m); Loading—IRC—CLASS—A; Safety Kerbs—60 cm

Materials; Concrete—M_{150}; Steel—All rolled sections, plates etc.

2·2. Deck Slab

Arrangement of cross-girders and stringer beams.
Span for the main girder=20 m
Let spacing of cross girders=4 m c/c
Spacing of stringer beams=1·7 m c/c
Span for deck slab=1·7 m
l'=width of slab=4 m
l=effective span of the slab=1·7 m

$$\frac{l'}{l}=\frac{4}{1\cdot 7}=2\cdot 35>2$$

Slab spans in one direction only. If a solid slab is supported on two opposite edges and carries concentrated loads, it should be designed to resist the maximum bending moment caused by the loading system. Such bending moment shall be assumed to be resisted by an effective width measured parallel to the supporting edges.

The effective width shall be calculated from the expression.

$$e=Kx\left(1-\frac{x}{l}\right)+W$$

wher K=a constant depending on $\frac{l'}{l}$ ratio,

x=distance of the C.G. of the concentrated load from the nearer suport.

W=tyre or track contact area over the road surface of the slab in the direction at right angles to the span + the thickness of the wearing coat or surface finish above the structural slab.

2·3. Loads

Dead loads

Weight of slab (20 cm)$=\frac{20}{100}\times 1\times 1\times 2400=480$ kg/m^2

Wearing coat (10 cm)$=\frac{10}{100}\times 1\times 1\times 2200=220$ kg/m²

Total dead load $=700$ kg/m²

Live loads. (Class-A loading.)

It is assumed that maximum bending moment is produced in the deck slab when the heaviest wheal load is at the centre of span.

Therefore $x=\frac{1\cdot7}{2}=0\cdot85$ m

Deck slab is continuous over stringer beams and for continuous slab $\frac{l'}{l}>2$

$K=2\cdot60$

$e=2\cdot6\times0\cdot85\left(1-\frac{\cdot85}{1\cdot7}\right)+0\cdot25+2\times0\cdot10=1\cdot555$ m

Effective width of two wheel loads overlap

Therefore the effective width

$=1\cdot2+\frac{1\cdot555}{2}+\frac{1\cdot555}{2}=2\cdot755$ m <4 m the width of the slab.

Wheel load per metre of slab$=\frac{5\cdot7\times2}{2\cdot755}=4\cdot12$ t

2·4. Impact on Live Loads

Impact factor for deck system upto 3 m span for steel bridges $=54\cdot5\%$

Live load per metre width including impact$=1\cdot545\times4\cdot12$ $=6\cdot34$ t

2·5. Bending Moments

(*i*) due to **D***L*

Maximum $-$ve $BM=\frac{w\,l^2}{10}=\frac{700\times1\cdot7^2}{10}=-205$ m kg

Maximum $+$ve $BM=\frac{w\,l^2}{12}=\frac{700\times1\cdot7^2}{12}=+168\cdot5$ m kg

(*ii*) due to Live loads

Load dispersion from the wheel in the direction of span $=0\cdot50+2(0\cdot1+0\cdot18)=1\cdot06$ m

Maximum moment at the centre of the span

$=\frac{6\cdot34}{2}\times\frac{1\cdot7}{2}-\frac{6\cdot34}{2}\times\frac{1\cdot06}{4}=2\cdot7-0\cdot84=1\cdot86$ mt

Due to continuity live load bending moment can be reduced.

Maximum $+$ve $BM=\frac{w\,l^2}{10}=\frac{8\times1\cdot86}{10}=+1\cdot49$ mt

Maximum $-$ve $BM=\frac{w\,l^2}{9}=\frac{1\cdot86\times8}{9}=-1\cdot655$ mt

2·6. Total Bending Moment

Total +ve BM due to $DL+LL$ including impact
$=16850+149000=165850$ cm kg
Total −ve BM due to $DL+LL$ including impact
$=20500+165500=186000$ cm kg.

2·7. Shear Force

(*i*) Due to $DL=0{\cdot}6\times wd\times l=0{\cdot}6\times 700\times 1{\cdot}7=714$ kg
(*ii*) Due to LL

Maximum shear Force occurs when the wheel loads are close to the supports as arranged.

2·8. Effective Width for wheel Load

(*i*) Load (1)

$$=K_1x_1\left(1-\frac{x_1}{L}\right)+W_1=2{\cdot}6\times 0{\cdot}53\left(1-\frac{0{\cdot}53}{1{\cdot}7}\right)+0{\cdot}45$$

$=1{\cdot}38\ (1-0{\cdot}312)+0{\cdot}45=0{\cdot}95+0{\cdot}45=1{\cdot}4$ m
Effective width$=1{\cdot}4+1{\cdot}2=2{\cdot}6$ m

(*ii*) Load (2)

$$e_2=K_2x_2\left(1-\frac{x_2}{L}\right)+W_2=2{\cdot}6\times 0{\cdot}63\left(1-\frac{0{\cdot}63}{1{\cdot}7}\right)+0{\cdot}45$$

$=1{\cdot}03+0{\cdot}45=1{\cdot}48$ m
Effective width$=1{\cdot}48+1{\cdot}2=2{\cdot}68$ m

(*iii*) Load (3)

$$e_3=K_3x_3\left(1-\frac{x_3}{2}\right)+W_3=2{\cdot}6\times 0{\cdot}35\left(1-\frac{0{\cdot}35}{1{\cdot}7}\right)+0{\cdot}45$$

$=0{\cdot}72+0{\cdot}45=1{\cdot}17$ m
Widths do not overlap, effective width$=1{\cdot}17$ m

Load/m width for load (1)$=\dfrac{2\times 5{\cdot}7}{2{\cdot}6}=4{\cdot}39$ t

Load/m width for load (2)$=\dfrac{2\times 5{\cdot}7}{2{\cdot}68}=4{\cdot}06$ t

Load/m width for load (3)$=\dfrac{5{\cdot}7}{1{\cdot}17}=4{\cdot}89$ t

Neglecting continuity
Maximum shear Force in span (1)

$$=4{\cdot}39\times\frac{(1{\cdot}7-053)}{1{\cdot}7}=3{\cdot}02\text{ t}$$

SF including impact on $LL=3{\cdot}02\times 1{\cdot}545=4{\cdot}66$ t
Maximum SF in span-2

$$=4{\cdot}39\times\frac{0{\cdot}53}{1{\cdot}7}+4{\cdot}06\,\frac{(1{\cdot}7-0{\cdot}63)}{1{\cdot}7}+4{\cdot}89\times\frac{0{\cdot}18}{1{\cdot}06}\times\frac{0{\cdot}18}{1{\cdot}7}$$

$=1{\cdot}37+2{\cdot}56+0{\cdot}0875=4{\cdot}0175$ t
SF including inpact on $LL=1{\cdot}545\times 4{\cdot}0175=6{\cdot}22$ t
Maximum SF due to $LL=6{\cdot}22$ t

Maximum SF due to all loads=0·714 t+6·22 t=6·934 t
=6934 kg

2·9. Design of Deck Slab

Maximum BM due to all loads=186000 cm kg
Maximum SF due to all loads=6934 kg
Using M_{150} concrete and grade-1 steel

$$d_e=\sqrt{\frac{186000}{100\times 8\cdot 7}}=14\cdot 6 \text{ cm}$$

Use d_e=15 cm and d=18 cm

2·9·1. Main Reinforcement

$$A_t=\frac{186000}{1400\times \cdot 87\times 15}=10\cdot 2 \text{ cm}^2;$$ Use 12 mm at 11 cm c/c (10·28 cm²).

2·9·2. Secondary Reinforcement

Secondary steel is provided to resist a bending moment
$=0\cdot 3\times LL$ moment $+0\cdot 2\times DL$ moment
=0·3×165500+0.2×20500=49650+4100=53750 cm kg
Using 8 mm dia. bars, effective depth available
=18−(2·5+1·2+0·4)=13·9 cm

$$A_t=\frac{53750}{1400\times \cdot 87\times 13\cdot 9}=3\cdot 16 \text{ cm}^2$$

Use 8 mm dia. at 15·5 cm c/c (3·24 cm²)

$$=\frac{S}{b\times a}=\frac{6934}{100\times \cdot 87\times 15}=5\cdot 3 \text{ kg/cm}^2>5 \text{ kg/cm}^2$$

Use fillets of 10 cm × 10 cm at the junctlon of the beam, rib or use d=20 cm, d_e=17 cm

2·10. Stringer Beam

Span=4 m
Dead load *BM*
Dead load due to self weight of slab and wearing coat
=700×1·7=1190 kg/m
DL due to stringer beams at 100 kg/m=100 kg/m
Total dead load=1290 kg/m

Max. *DL* bending moment $=\frac{1290\times 4\times 4\times 100}{8}$ =258,000 cm kg

Maximum SF $=\frac{1290\times 4}{2}$ =2580 cm kg

2·10·1. Live load BM

Live load on stringer beam (2)

$$=W\,\frac{0\cdot 4}{1\cdot 7}+W\,\frac{1\cdot 2}{1\cdot 7}=0\cdot 2365\,W+0\cdot 706\,W=0\cdot 9425\,W$$

Live load on stringer (3)

$$=W\,\frac{0\cdot 5}{1\cdot 7}+W\,\frac{1\cdot 48}{1\cdot 7}=0\cdot 294W+0\cdot 871\,W=1\cdot 165\,W$$

Live load on stringer (4)

$$=W\times\frac{0\cdot22}{1\cdot7}+W\frac{1\cdot38}{1\cdot7}=(0\cdot1295+0\cdot812)W=0.9415\ W$$

Maximum bending moment occurs under the load having the maximum axle load (i.e. 11·4 t) when C.G. of the loads and the load under which the *BM* is maximum are equidistant from the centre of the span.

Maximum wheel load$=W=5\cdot7$ t

$$\text{Maximum } BM=\frac{5\cdot7\times1\cdot165\times2\times1\cdot7\times1\cdot7}{4}=9\cdot65 \text{ mt}$$

$$=965000 \text{ cm kg}$$

2·10·2. SF Due to LL

$$\text{Maximum SF}=6\cdot65\frac{(4-0\cdot405)}{4}+6\cdot65\frac{(4-1\cdot1605)}{4}$$

$$=6\cdot65\times\frac{3\cdot595}{4}+6\cdot65\times\frac{2\cdot395}{4}=\frac{39\cdot75}{4}=9\cdot935 \text{ t}$$

2·11. Impact Factor

Impact factor for steel bridges for class *A* loading

$$=\frac{9}{13\cdot5+L}=\frac{9}{13\cdot5+4}=0\cdot514$$

Max. *BM* due to *LL* including impact$=1\cdot514\times9\cdot65=14\cdot6$ mt

Max. *SF* due to *LL* including impact$=1\cdot514\times9\cdot935=15\cdot04$ t

2·12. Design of Stringer

Maximum *BM* due to $DL+LL$ including impact$=2\cdot58+14\cdot6$ $=17\cdot18$ mt

Maximum *SF* due to all loads$=2\cdot58+15\cdot04=17\cdot62$ t

$$\text{Modulus of section required}=\frac{1718000}{1500}=1140 \text{ cm}^3$$

Use ISLB$-$450, $Z_{xx}=1223\cdot8$ cm^3

2·13. Shear Stress

Shear Force$=17620$ kg

$$\text{Shear stress}=\frac{17620}{45\times0\cdot86}=455 \text{ kg/cm}^2<945 \text{ kg/cm}^2$$

2·14. Cross Girder

Span$-$8 m

DL due to slab and wearing coat$=700\times4=2800$ kg/m

This load is assumed to be uniformly distributed.

Dead load due to stringer beam$=65\cdot3\times4=251\cdot2$ kg

Say$=260$ kg

Self weight of cross girder Say$=200$ kg/m

Total udl$=2800+200=3000$ kg/m$=3$ t/m

2·14·1. BM due to Dead Loads

$$R_A=\tfrac{1}{2}\ (3\times8+5\times0\cdot26)=\frac{25\cdot3}{2}=12\cdot65 \text{ t}$$

Max. $BM = 12{\cdot}65 \times 4 - 3 \times 4 \times \frac{4}{2} - 0{\cdot}26 \times 3{\cdot}4 - 0{\cdot}26 \times 1{\cdot}7$
$= 50{\cdot}60 - 24 - 0{\cdot}885 - {\cdot}442 = 25{\cdot}273$ mt

Max. $SF = 12{\cdot}65$ t

2·14·2. Live Load Bending Moment

Arrangement of live loads for class *A* loading to obtain maximum reaction on the cross girder

1st trial arrangement.

Neglecting continuity due to other spans

Reaction R_1 due to one set of wheel load for the arrangement

$$R_1 = 5{\cdot}7 + 1{\cdot}35 \times \frac{0{\cdot}8}{4} + 5{\cdot}7 \times \frac{2{\cdot}8}{4} = 5{\cdot}7 + 0{\cdot}27 + 3{\cdot}99 = 9\ 96 \text{ t}$$

2nd trial arrangement:

$$R_2 = 5{\cdot}7 + 5{\cdot}7 \times \frac{2{\cdot}8}{4} = 5{\cdot}7 + 3{\cdot}99 = 9{\cdot}69 \text{ t}$$

The first trial arrangement produces max. *BM* in the cross girder.

The arrangement of loads to produce maximum *BM* in the cross girder is to arrange them symmetrically from centre of the span.

Max. $BM = R \times 4 - 9{\cdot}96 \times (4 - 1{\cdot}49) - 9{\cdot}96\ (4 - 3{\cdot}29)$
$= 9.96 \times 2 \times 4 - 9.96 \times 2{\cdot}51 - 9{\cdot}96 \times 0{\cdot}71 = 79{\cdot}8 - 25 - 7{\cdot}1$
$= 47{\cdot}7$ mt

2·14·3. Impact Factor on LL

Impact factor for closs-A loading

$$= \frac{9}{13{\cdot}5 + L} = \frac{9}{13{\cdot}5 + 8} = 0{\cdot}42$$

BM due to *LL* including impact $= 1{\cdot}42 \times 47{\cdot}7 = 68$ mt

2·14·4. Shear Force

Maximum shear force will occur when the wheel load is close to the support.

$R_A \times 8 = 9{\cdot}96[(8-1) + [(8 - 2{\cdot}8) + (8 - 4{\cdot}22) + (8 - 6{\cdot}02)]$

$$R_A = \frac{9{\cdot}96 \times 17{\cdot}96}{8} = 22{\cdot}4 \text{ t}$$

Max. *SF* including impact $= 22{\cdot}4 \times 1{\cdot}42 = 31{\cdot}8$ t

2·15. Design of Girder

Span $= 8$ m

Total bending moment due to *DL, LL* including impact
$= 25{\cdot}273 + 68 = 93{\cdot}273$ mt

Total *SF* due to *DL, LL* including impact $= 12{\cdot}65 + 31{\cdot}80$
$= 44{\cdot}55$ t

$$\text{Modulus of secton required} = \frac{9327300}{1500} = 6220 \text{ cm}^3$$

ISLB $-$ 600 provides $Z_{xx} = 2428{\cdot}9$ cm^3 only

Using welded composite beam

Area of one cover plate required in composite beam

$$=\frac{6220-2428\cdot9}{60}=\frac{3791\cdot1}{60}=63 \text{ cm}^2$$

Use 45 × 1·5 cm cover plates.

2·15·1. Shear Stress

$$V_s=\frac{S}{dt_w}=\frac{44450}{60\times1\cdot05}=705 \text{ kg/cm}^2<945 \text{ kg/cm}^2$$

2·16. Main Girder

Span = 20 m

2·16·1. Dead Loads

Dead load reaction from cross girder to main girder at 4 m c/c = 12·65 t

Self weight of main girder $w=20\,L+100$ kg/m $=20\times20+100$ = 500 kg/m

2·16·2. Live Loads

The effect of live loads due to IRC class 'A' loading will be maximum when the vehicle is closest to the kerb. When the vehicle moves closest to kerb load transferred to main girder due to any wheel load $W=\frac{W}{8}\,(7+5\cdot2+3\cdot78+1\cdot98)=2\cdot24\,W$

Impact factor for 20 m span main girder on live loads

$$=\frac{9}{13\cdot5+L}=\frac{9}{13\cdot5+20}=\frac{9}{33\cdot5}=0\cdot269$$

Live load including impact = $1\cdot269\times2\cdot24\,W=2\cdot84\,W$

All the wheel loads are multiplied by this factor 2·84 and these modified wheel loads are considered for computing *BM* and *SF* on the main girder.

2·16·3. Bending Moments Live Loads

BM is maximum at the centre of the span when

$\frac{W_1}{L_1}-\frac{W_2}{L_2}$ changes sign.

Try load (4) at centre of span. This load satisfies the condition From Influence line for *BM* at centre of span.

$$\text{IL ordinate at cente}=\frac{x(l-x)}{l}=\frac{10\times(20-10)}{20}=\frac{100}{20}=5$$

BM due to rolling loads

$$=\frac{5}{10}\,(4\cdot5\times3\cdot84+5\cdot6\times3\cdot84+8\cdot8\times16\cdot2+16\cdot2\times10$$
$$+5\cdot7\times9\cdot65+2\cdot7\times9\cdot65)$$

$$=\frac{5}{10}\times424\cdot9=212 \text{ mt.}$$

2·16·4. BM due to DL

BM due to loads from cross girders to main girder at centre of span$=25{\cdot}30\times10-12{\cdot}65\times6-12{\cdot}65\times2=253-75{\cdot}90-25{\cdot}3$
$=151{\cdot}8$ mt.

2·16·5. BM due to Self Weight

BM due to self weight of girder$=\dfrac{wl^2}{8}=\dfrac{0.5\times20\times20}{8}$
$=25$ mt

BM due to all loads$=212+151{\cdot}8+25=388{\cdot}8$ mt.

2·17. Shear Force Live Loads

SF due to live loads is maximum when load (3) is on the support. From the influence line for reaction at support.

Maximum reaction at support$=16{\cdot}2\times1+\dfrac{18{\cdot}8}{20}\times16{\cdot}2$

$+\dfrac{9{\cdot}65}{20}(14{\cdot}5+11{\cdot}5+8{\cdot}5+8{\cdot}5)=50{\cdot}71.$

2·17·1. Dead Loads

SF due to reactions from cross girders$=\dfrac{\text{Total load}}{2}$

$=\dfrac{12{\cdot}65\times5}{2}=31{\cdot}675$ t

SF due to self weight of girder$=\dfrac{0{\cdot}5\times20}{2}=5$ t

SF due to all loads$=50{\cdot}71+31{\cdot}675+5=86{\cdot}385$ t

2.18. Plate Girder Design

Economical depth$=5\sqrt[3]{\dfrac{M}{f_b}}=5\sqrt[3]{\dfrac{3888000}{1500}}$

$=5\sqrt[3]{26000}=29{\cdot}6\times5=148$ cm.

Use $d=150$ cm

2·18·1. Web Plate

Web thickness required to resist shear$=\dfrac{86385}{945\times150}$

$=0{\cdot}61$ cm$=6{\cdot}1$ mm

Use minimum thickness of 8 mm for highway bridges. Use web plate 1500×8mm.

2·18·2. Flange Plates

$$M=f_b\left(A_f+\frac{A_w}{6}\right)d$$

$$A_f=\frac{38{\cdot}88\times10^6}{150\times1500}-\frac{150\times8}{6}=173-20=153\text{ cm}^2$$

Width of flange$=\frac{L}{40}=\frac{20\times 100}{40}=50$ cm

Thickness of flange plate required$=\frac{153}{50}=3{\cdot}06$ cm. Use two 20 mm plates. Use 500×40 mm flange plates.

2·19. Stresses

MI of the section$=\frac{1}{12}\times 0{\cdot}8\times(150)^3+2\times 59\times 4\times(77)^2$
$=22550+2368000=2{\cdot}39\times 10^6$ cm^4

$$f_b=\frac{M_y}{I}=\frac{38{\cdot}88\times 10^6\times 79}{2.39\times 10^6}=1290 \text{ kg/cm}^2<1500 \text{ kg/cm}^2.$$

2·20. Connection

$$v=\frac{F A \bar{y}}{I}\frac{86385\times 50\times 4\times 77}{2{\cdot}39\times 10^6}=555 \text{ kg/cm}$$

Assuming continuous weld on either side of the web plate, weld size ×0·7×1025×2=555

$$=\frac{555}{0{\cdot}7\times 2\times 1025}=0{\cdot}385 \text{ cm}=3{\cdot}85 \text{ mm}$$

Maximum size of the weld=2/3×0 8=5 mm
Use 5 mm continuous weld on either side of the web

2·21. Stiffners

$$\frac{d}{t_w}=\frac{150}{0{\cdot}8}=187{\cdot}5<200$$

No horizontal stiffners are needed.

Since $\frac{d}{t_w}>85$ Intermediate stiffness are needed

2·21·1. Intermediate Stiffner

Spacing of stiffner $\frac{d_w}{3}$ to $1{\cdot}5\ d_w=\frac{150}{3}$ to $1{\cdot}5\times 150=50$ cm to 225 cm
Try stiffners at 100 cm c/c.

2·21·2. Panel Dimensions

$180\ t_w=180\times 0{\cdot}8=144$ cm
$270\ t_w=270\times 0{\cdot}8=220$ cm

M.I. for the stiffner$=\frac{1{\cdot}5\times 150^3\times 0{\cdot}8^3}{100^2}=260$ cm^4

Try 100×8 mm stiffner.
MI about base for the stiffner$=1/3\times 0{\cdot}8\times 10^3=266{\cdot}5$ cm^4 >260 cm^4.

2·21·3. Connection

Shear to be resisted$=\frac{t^2}{2h}=\frac{(0{\cdot}8)^2}{2\times 10}=0{\cdot}032$ t/cm

Size of the weld$=\frac{{\cdot}032\times 1000}{{\cdot}7\times 1025}=0{\cdot}039$ cm$={\cdot}39$ mm

Use minimum size = 3 mm. Length of weld = $10 \times 0{\cdot}8 = 8$ cm.

Use 10 cm long 3 mm size weld staggered on either side of the stiffner.

2·21·4. Bearing Stiffners

Load on stiffner = end reaction = 86·385 t

$$\text{Bearing area required} = \frac{86385}{1890} = 45{\cdot}5 \text{ cm}^2.$$

Effective length of stiffner as column = $0{\cdot}7 \times 150 = 105$ cm

Try 240×20 mm plates as stiffners on either side of the web

$I_{xx} = \frac{1}{2} \times 2 \times (48{\cdot}8)^3 + 1/12 \times 22(0{\cdot}8)^3 = 19400 \text{ cm}^4$

$A = 48{\cdot}8 \times 2 + 22 \times 0{\cdot}8 = 115{\cdot}2 \text{ cm}^2$

$$xx = \sqrt{\frac{I}{A}} = \sqrt{\frac{19400}{115{\cdot}2}} = 12{\cdot}95 \text{ cm}$$

$$\frac{l_e}{r_{xx}} = \frac{105}{12{\cdot}95} = 8{\cdot}12$$

allowable stress $f_e = 1246{\cdot}7$ kg/cm²

$$\text{Required area} = \frac{86385}{1246{\cdot}7} = 69 \text{ cm}^2 < 115{\cdot}2 \text{ cm}^2.$$

2·21·5. Connection

Length of web available for connection using intermitent staggered welds = $2 \times 150 = 300$ cm

$$\text{Shear to be resisted} = \frac{86385}{300} = 287.9 \text{ kg/cm}$$

$$\text{Size of the weld} = \frac{287{\cdot}9}{0{\cdot}7 \times 1025} = {\cdot}4 \text{ cm} = 4 \text{ mm}$$

Use 4 mm weld of 75 mm length intermitent on either side of the web.

DESIGN OF PLATE GIRDER TWO LANE HIGHWAY BRIDGE SPAN 20M

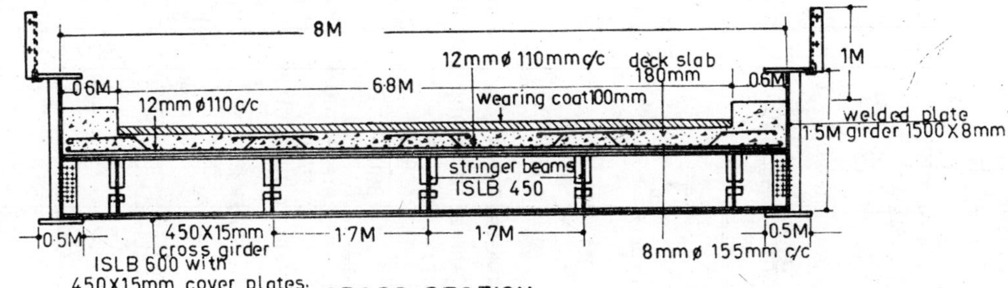

CROSS SECTION

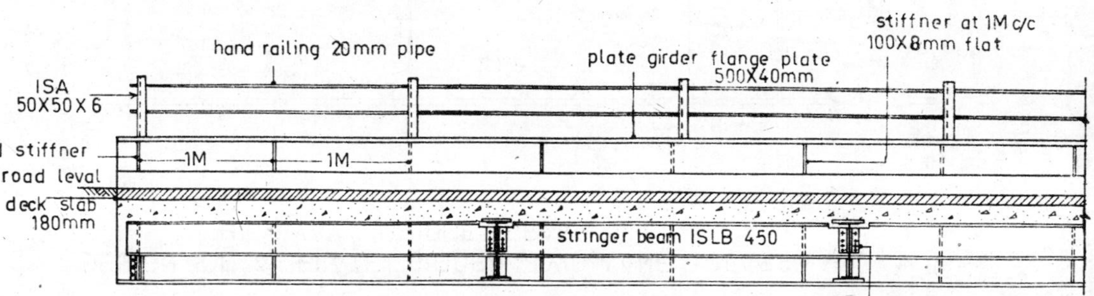

LONGITUDINAL SECTION

DESIGN OF A PLATE GIRDER TWO LANE CLASS-A HIGHWAY BRIDGE · SPAN - 20M

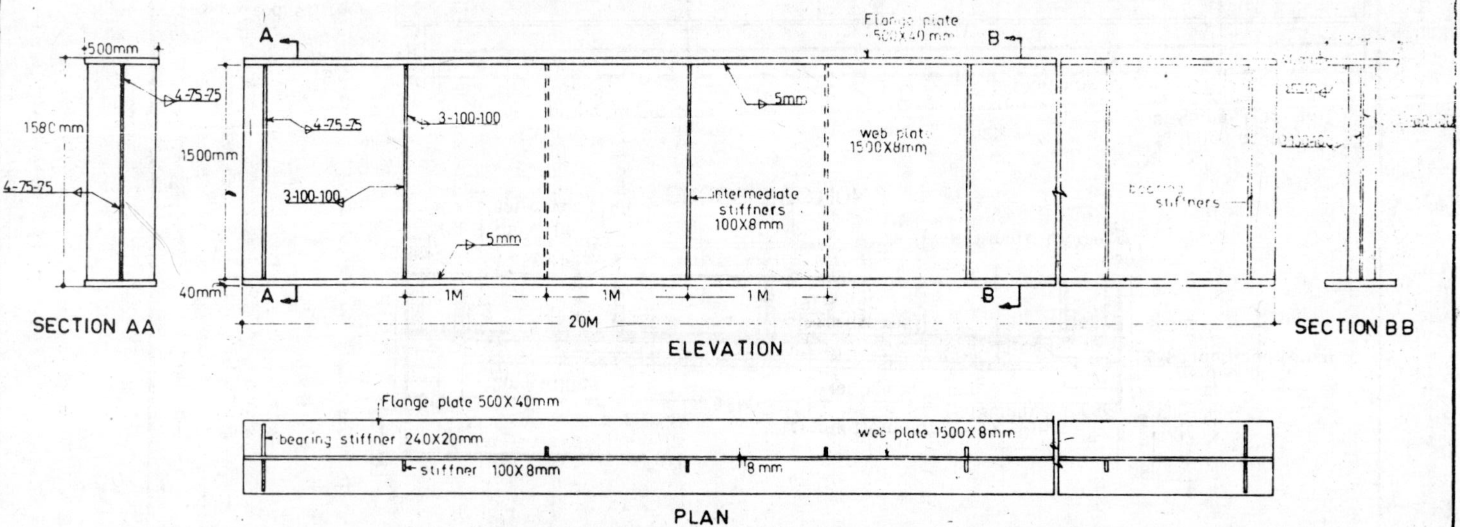

DETAILS OF WELDED PLATE GIRDER

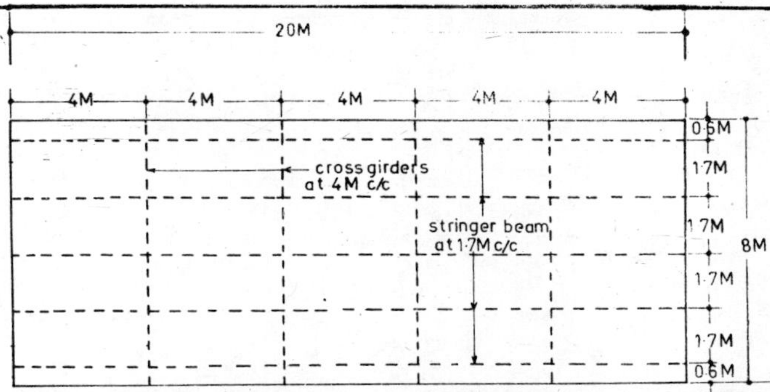

PLAN SHOWING THE ARRANGEMENT OF CROSS GIRDERS & STRINGERS

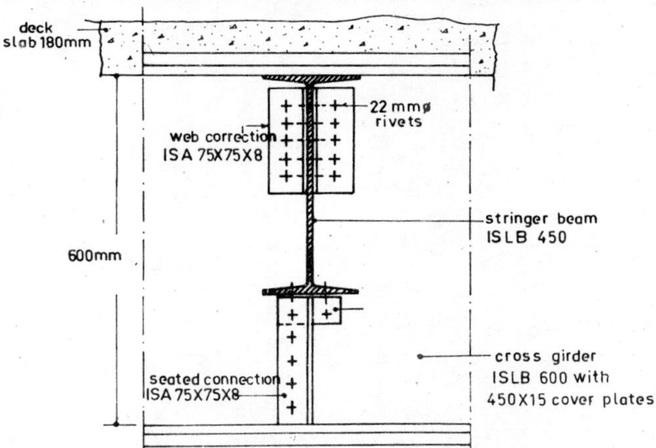

STRINGER BEAM CONNECTION

WELDED PLATE GIRDER CROSS GIRDER CONNECTION

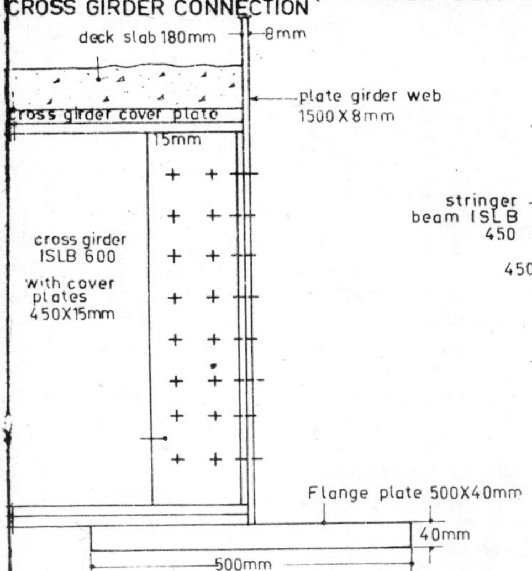

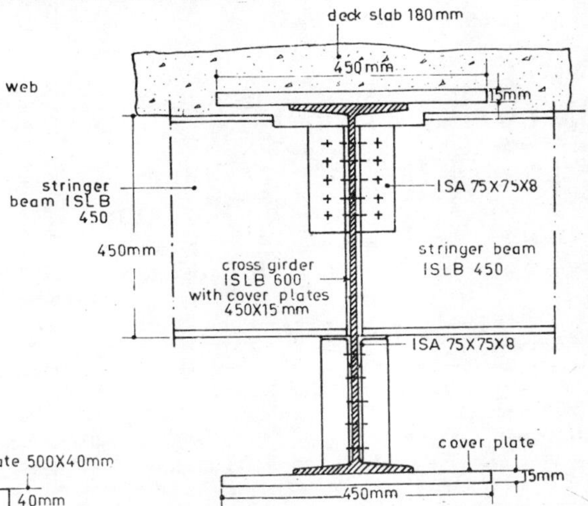

CROSS GIRDER – STRINGER BEAM

3. Design of Steel Trussed Girder 54 M Span—Highway Bridge

3·1. Data

Type—Through type; Roadway—Two lane 6—8 m;
Loading—IRC Class—A; Safety Kerbs—60 cm;
Material available; Concrete—M_{150}; Steel—All rolled sections.

3·2. Deck Slab

Arrangement of cross girders and stringer beams span for the main girder = 54 m

Let cross girder be placed at 6 m centre to centre and the stringer beams at 2·25 m c/c

Span for deck slab = 2·25 m

l' = width of the slab

l = effective span of the slab

$$\frac{l'}{l} = \frac{6}{2 \cdot 25} = 2 \cdot 67 > 2.$$

Slab spans in one direction only. If a solid slab is supported on two opposite edges and carries concentrated loads, it should be designed to resist the maximum bending moment caused by the loading system. Such bending moment shall be assumed to be resisted by an effective width measured parallel to the supporting edges.

The effective width shall be calculated from the expression

$$e = Kx\left(1 - \frac{x}{l}\right) + W$$

where K = a constant depending on $\frac{l'}{l}$ ratio

x = distance of the C.G. of the concentrated load from the nearer support

W = Tyre or track contact area over the road surface of the slab in the direction at right angles to the span + twice the thickness of the wearing coat or surface finish above the structural slab.

3.3. Loads

Dead loads

$$\text{Weight of slab (20 cm)} = \frac{20}{100} \times 1 \times 1 \times 2400$$

$$= 480 \text{ kg/m}^2$$

Weight of coat (10 cm) $=\frac{10}{100}\times 1\times 2200$

$=220\ kg/m^2$

Total dead load $=700\ kg/m^2$.

Live loads

Class A loading system consists of

↓ 1·1m ↓ 3·2m ↓ 1·2m ↓ 4·3m ↓ 3m ↓ 3m ↓ 3m ↓

2·7 t 2·7 t 11·4 t 11·4 t 6·8 t 6·8 t 6·8 t 6·8 t.

It is assumed that maximum bending moment is produced in the deck slab when the heaviest wheel load is at the centre of span.

Therefore the distance from the nearer support to the C.G. of the concentrated load

$$x=\frac{2{\cdot}25}{2}=1{\cdot}125\ m$$

Effective width of the slab in the direction of travel of the wheel load considered

$$e=Kx\left(1-\frac{x}{l}\right)+W$$

For continuous slabs $\frac{l'}{l}>2$ from Table

$K=2{\cdot}60$

$$e=2{\cdot}6\times 1{\cdot}125\left(1-\frac{1{\cdot}125}{2{\cdot}25}\right)+{\cdot}25+2\times 0{\cdot}10=1{\cdot}91\ m$$

Effective width of two wheel loads overlap

Therefore the effective width $=1{\cdot}2+\frac{1{\cdot}91}{2}+\frac{1{\cdot}91}{2}$

$=3{\cdot}11\ m<6m$, the width of slab.

Wheel load per metre of slab $=\frac{2\times 5{\cdot}7}{3{\cdot}11}=3{\cdot}65\ t$

3·4. Impact on Live Load

Impact factor for deck system upto 3 m for steel bridges $=54{\cdot}5\%$

Live load per metre width including impact $=3{\cdot}65\times 1{\cdot}545$

$=5{\cdot}65\ t$

3·5. Bending Moments

Dead load bending moments

Maximum—*ve* BM $=\frac{wl^2}{10}=\frac{700\times 2{\cdot}25^2\times 100}{10}$

$=-35500$ cm kg

Maximum+*ve* BM $=\frac{wl^2}{12}=\frac{700\times 2{\cdot}25^2\times 100}{12}$

$=+29{,}500$ cm kg

Live load bending moment

Load dispersion from the wheel in the direction of span
$=0{\cdot}50+2({\cdot}10+{\cdot}18)=1{\cdot}06$ m
Miximum bending moment at the centre of the span

$$=\frac{5{\cdot}65}{2}\times\frac{2{\cdot}75}{2}-\frac{5{\cdot}65}{2}\times\frac{1{\cdot}06}{4}=\frac{5{\cdot}65}{2}(1{\cdot}125-0{\cdot}265)$$

$$=\frac{5{\cdot}65}{2}\times{\cdot}86=2{\cdot}42 \text{ mt}=242{,}000 \text{ cm kg}$$

Due to continuity live load bending moments can be reduced.

$$\text{Maximum}+ve \text{ bending moment}=\frac{wl^2}{10}=0{\cdot}8\times242000$$
$$=197{,}000 \text{ cm kg}$$

$$\text{Maximum}-ve \text{ bending moment}=-\frac{wl^2}{9}=-\frac{8}{9}\times242000$$
$$=-215{,}000 \text{ cm kg.}$$

3·5·1. Total Bending Moment

Total $+ve$ bending moment due to DL+LL including impact
$=29{,}500+197{,}000=226{,}400$ cm kg
Total $-ve$ BM due to DL+LL including impact
$=35{,}500+215{,}000=250{,}500$ cm kg
Maximum BM=250,500 cm kg.

3·6. Shear Force

Due to $\text{DL}=0{\cdot}6\times wd\times l=0{\cdot}6\times700\times2{\cdot}25=945$ kg

Maximum shear force occurs when the wheel loads are close to supports as arranged.

3·6·1. Effective width for Wheel Loads (1)

Effective width for wheel load (1)

$$e_1=Kx_1\left(1-\frac{x_1}{l}\right)+W$$

$x_1=0{\cdot}53$

$$e_1=2{\cdot}6\times0{\cdot}53\left(1-\frac{0{\cdot}53}{2{\cdot}25}\right)+0{\cdot}45=1{\cdot}51 \text{ m}$$

Since the effective widths overlap for the loads
total effective width $=1{\cdot}51+1{\cdot}20=2{\cdot}71$m.

3·6·2. Effective width for Wheel Loads (2)

Effective width $x_2=0{\cdot}08$

$$e_2=2{\cdot}6\times0{\cdot}8\left(1-\frac{{\cdot}08}{2{\cdot}75}\right)+{\cdot}45=0{\cdot}65 \text{ m}$$

The effective widths do not overlap
The effective width $=0{\cdot}65$ m
Loads per metre width due to wheel loads

$$(1)\ =\frac{5{\cdot}7\times2}{2{\cdot}71}=4{\cdot}2 \text{ t}$$

Load per metre width due to wheel loads

$$(2)\ =\frac{5{\cdot}7}{0{\cdot}65}=8{\cdot}8 \text{ t}$$

Neglecting continuity SF due LL

$$=R_A=4{\cdot}2\left(\frac{2{\cdot}25-0{\cdot}53}{2{\cdot}25}\right)+\left(\frac{8{\cdot}8}{1{\cdot}06}\times0{\cdot}45\right)\times\frac{0{\cdot}45}{2\times2{\cdot}25}$$

$=3{\cdot}2+0{\cdot}375=3{\cdot}575$ t

Impact on live load $=0{\cdot}545$

SF due to LL including impact $=1{\cdot}545\times3{\cdot}575=5{\cdot}525$ t

Maximum SF due to DL and LL including impact

$=945+5525=6470$ kg

3·7. Design of Deck Slab

Maximum BM due to all loads $=250{,}500$ cm kg

Maximum SF due to all loads $=6470$ kg

Using M_{150} concrete

$$8{\cdot}7\ bd^2=250{,}500$$

$$d^2=\frac{250{,}500}{8{\cdot}7\times100}\quad d=20\text{ cm}$$

$$d_e=\sqrt{\frac{250500}{8{\cdot}7\times100}}=17\text{ cm}$$

cover ≮ 2·5 cm; Use $d=20$ cm $d_e=17$ cm.

3·7·1. Main Reinforcement

$$A_t=\frac{250{,}500}{1400\times.87\times17}=10{\cdot}55\text{ cm}^2$$

Use 12 mm ϕ at 10 cm c/c (11·31 cm²) $<2\times$ effective depth $=34$ cm or 30 cm.

3·7·2. Secondary Reinforcement

Secondary steel is provided to resist a bending moment equal to $0{\cdot}3\times$ Live load moment $+0{\cdot}2$ times the dead load moment

$=0{\cdot}3\times215000+0{\cdot}2\times35500=64500+7100=71{,}600$ cm kg

Using 10 mm ϕ bars effective depth available

$=20-(2{\cdot}5+1{\cdot}2+0{\cdot}5)=15{\cdot}8$ cm

$$A_t=\frac{71400}{1400\times{\cdot}87\times15{\cdot}8}=3{\cdot}72\text{ cm}^2$$

Use 10 mm ϕ at 20 cm c/c (3·93 cm²) <3 effective depth $3\times17=51$ cm or 45 cm

3·8. Shear Stress

$$q_s=\frac{s}{b\times a}=\frac{6470}{100\times{\cdot}87\times17}=4{\cdot}38\text{ kg/cm}^2<5\text{ kg/cm}^2$$

3·8·1. Effective width for wheel loads (3)

Load spread for load (3)

$$x=0{\cdot}75\text{ m}=2{\cdot}6\times0{\cdot}75\left(1-\frac{0{\cdot}75}{2{\cdot}25}\right)+0{\cdot}45=1{\cdot}29+0{\cdot}45=1{\cdot}74\text{ m}$$

effective widths overlap

∴ effective width $=1{\cdot}2+1{\cdot}74=2{\cdot}94$ m

$$\text{Load per metre width}=\frac{5{\cdot}7\times2}{2{\cdot}94}=3{\cdot}88\text{ t}$$

3·8·2. Shear in the second span

For the position of wheel loads neglecting continuity at the support

$$\text{Maximum shear force} = \frac{8{\cdot}8 \times 0{\cdot}61}{1{\cdot}06} \times \frac{2{\cdot}17}{2{\cdot}25} + 3{\cdot}88 \times \frac{0{\cdot}75}{2{\cdot}95}$$

$$= 4{\cdot}86 + 1{\cdot}29 = 6{\cdot}15 \text{ t}$$

Shear force including impact $= 1{\cdot}545 \times 6150 = 9500$ kg
Shear force due to DL $= 945$ kg
Total shear force $= 10445$ kg

3·8·3. Shear Stress

$$\text{Shear stress} = \frac{10{,}445}{100 \times {\cdot}87 \times 17} = 7{\cdot}1 \text{ kg/cm}^2 > 5 \text{ kg/cm}^2$$

Slab fails in shear . Therefore Use M_{200} concrete or increase the depth at supports by providing fillets

By providing fillets of 15 cm × 15 cm
depth available at the edge or support $= 20 + 15 = 35$ cm
effective depth $= 32$ cm

$$\text{shear stress} = \frac{10{,}445}{100 \times .87 \times 32} = 3{\cdot}75 \text{ kg/cm}^2 < 5 \text{ kg/cm}^2$$

3·9. Stringer beam dead loads

Dead load due to self weight of slab and wearing coat
$= 700 \times 2{\cdot}25 = 1575$ kg/m
Dead load due to stringer (assume) $= 100$ kg/m
$\left(\frac{1}{10} \text{ to } \frac{1}{12} \text{ of span}\right)$

Total dead load on the stringer $= 1675$ kg/m

$$\text{Maximum } BM \text{ due to deam load} = \frac{1675 \times 6 \times 6 \times 100}{8}$$

$$= 755.000 \text{ cm kg}$$

$$\text{Shear force due to } DL = \frac{1675 \times 6}{2} = 5025 \text{ kg}$$

3·9·1. Live load

To produce the maximum effect in the beam, the arrangement is as shown

$$\text{Live load on stringer (2)} = \frac{w(2{\cdot}25 - {\cdot}075)}{2{\cdot}25} + \frac{W \times 0{\cdot}905}{2{\cdot}25}$$

$$+ \frac{W \times 0{\cdot}375}{2{\cdot}25} = 1{\cdot}55 \text{ W}$$

3·9·2. Bending moment due to L. L

Maximum bending moment occurs under the load (1) when C. G of the loads and the load (1) are equidistant from the centre of the span W=5·7 t maximum wheel load

$$\text{Maximum } BM = 5{\cdot}7 \times 1{\cdot}55 \times 2 \times \frac{2{\cdot}7}{6} \times 2{\cdot}7 = 21{\cdot}4 \text{ mt}$$

$$= 2140{,}000 \text{ cm kg}$$

3 9·3. Shear force due to L. L

The maximum SF occurs for the load position
Magnitudes of loads (1) and (2)=1·55×5·7=8·85 t
Magnitude of load (2)=1·55×3·4=5·3 t
Load spread for loads (1) and (2)=25+2×28=81 cm
Load spread for loads (3)=20+2×28=76 cm
Maximum SF

$$=8{\cdot}85\left(\frac{6-{\cdot}405}{6}\right)+8{\cdot}85\times\frac{4{\cdot}345}{6}+\frac{5{\cdot}3\times0{\cdot}045}{6}$$

=8·25+6·36+·0396=14·3496 t =14·35 t

3·9·4. Impact factor on LL

Impact factor for steel bridges for class—A loading

$$=\frac{9}{13{\cdot}5+L}=\frac{9}{13{\cdot}5+6}=0{\cdot}46$$

Maximum bending moment due to LL including impact =1·46×21·4=31·2 mt

Maximum SF including impact=1·46×14·35=20·8 t

3·10. Design of Stringer

Maximum *BM* due to *DL* + *LL* including impact=755000+ 3120000=3875,000 cm kg

Maximum *SF* due to *DL* + *LL* including impact=5025+20800 25825 kg

$$\text{Modulous of section required}=\frac{3875000}{1500}=2580 \text{ cm}^3$$

Maximum section *I SLB*—600 is not sufficient

Try *ISLB*—450×170 with two cover plates Z_{xx} available =1223·8 cm³

$$\text{Area of fillet required}=\frac{2580-1223{\cdot}8}{45}=\frac{1356{\cdot}2}{45}=30 \text{ cm}^2$$

Use 30×1 cm plates

$$\text{Shear stress}=\frac{25825}{45\times{\cdot}86}=675 \text{ kg/cm}^2 < 945 \text{ kg/cm}^2$$

Dead lock check

weight of *I SLB* 450	=65·3 kg/m
weight of two plates of 30×1 cm	=25·0 kg/m
Total	=90·3 kg/m<100 kg/m assumed

3·11. Cross girders

Dead loads

Span of cross girder =8 m
Dead load due to slab and wearing coat=700×6=4200 kg/m
This is assumed to be uniformly distributed.
Dead load due to stringer=90 25×6=541·8 kg
Add 10% for connections =54·1 kg
Total =595·9 kg say=600 kg

This load is assumed to be concentrated

Self weight of the girder (assume) $(20\,L+100)$ kg/m $=300$ kg/m

Total v. d. L on the beam $=4500$ kg/m

3·11·1. Bending due to DL

$$\text{Maximum } BM \text{ at centre} = R_A\times4-4500\times4\times\frac{4}{2}-600(4-{\cdot}625)$$
$$-600\,(4-2{\cdot}875)$$

$$R_A=\tfrac{1}{2}[4500\times8+4\times600] \quad =\tfrac{1}{2}(36000+2400)$$
$$=\frac{38400}{2}=19200 \text{ kg}$$

$$BM=19200\times4-37600-2020-680=76800-40300$$
$$=36500 \text{ m kg}=3650{,}000 \text{ cm kg}$$

Maximum SF due to dead load $=19200$ kg

3·11·2. Live loads

Arrangement of live loads for class-A loading to obtain maximum reaction on the cross-girder.

1st trial arrangement. Neglecting conitnuity due to other spans.

Reaction R_1 due to one set of wheel loads for the arrangement shown

$$=5{\cdot}7+1{\cdot}35\times\frac{1{\cdot}7}{6}+1{\cdot}35\times\frac{2{\cdot}8}{6}+5{\cdot}7\times\frac{4{\cdot}8}{6}+3{\cdot}4\times\frac{0{\cdot}5}{6}$$
$$=5{\cdot}7+0{\cdot}63+0{\cdot}382+4{\cdot}56+0{\cdot}284=11{\cdot}556 \text{ t}$$

2nd trial arrangement

$$R_2=5{\cdot}7+1{\cdot}35\times\frac{0{\cdot}5}{6}+1{\cdot}35\times\frac{1{\cdot}5}{6}+\frac{5{\cdot}7\times4{\cdot}8}{6}+3{\cdot}4\times\frac{1{\cdot}7}{6}$$
$$=5{\cdot}7+0{\cdot}112+0{\cdot}36+4{\cdot}56+0{\cdot}962=11{\cdot}694 \text{ t say } 11{\cdot}7 \text{ t}$$

3·11·3. Bending Moment Due to LL

The arrangement of loads to produce maximum BM in the cross girder is to arrange them symmetrically from centre of the span

Maximum BM for the arrngement

$$=R\times4-11{\cdot}7\times(4-1{\cdot}49)-11{\cdot}7\,(4-3{\cdot}29)$$

$$R=\frac{11{\cdot}7\times4}{2}=23{\cdot}4 \text{ t}$$

$$\text{BM}=23{\cdot}4\times4-11{\cdot}7\times2{\cdot}51-11{\cdot}7\times0{\cdot}71=56 \text{ mt}$$
$$=5600{,}000 \text{ cm kg}$$

3·11·4. Impact Factor on LL

Impact factor for class—A loading

$$=\frac{9}{13{\cdot}5+L}=\frac{9}{13{\cdot}5+8}=0{\cdot}42$$

BM due to LL including impact

$$=1{\cdot}42\times5600{,}000=7950{,}000 \text{ cm kg}$$

3·12. Shear Force

Maximum shear force will occur when the wheel load is close to the support

$$R_A \times 8 = 11{\cdot}7[(8-1)+(8-2{\cdot}8)+(8-4{\cdot}22)+(0-6{\cdot}02)]$$
$$= 11{\cdot}7 \times 17{\cdot}96$$

$$R_A = \frac{11{\cdot}7 \times 17{\cdot}96}{8} = 26 \text{ t} = 26000 \text{ kg}$$

Maximum shear force including impact $= 26000 \times 1{\cdot}42$
$= 37000$ kg

3·13. Design of Cross Girder

Span $= 8$ m

Total bending moment due to DL, LL including impact
$= 3650{,}000 + 7950{,}000 = 11600{,}000$ cm kg

Total SF due to DL, including impact
$= 19200 + 37000 = 56{,}200$ kg

$$\text{Modulus of section required} = \frac{11600{,}000}{1500} = 7720 \text{ cm}^3$$

Maximum size rolled steel section provided only $z = 2428{\cdot}9$ cm³

Using welded connections

Area of one cover plate required in composite beam

$$= \frac{7720-2428{\cdot}9}{60} = \frac{5291{\cdot}1}{60} = 89{\cdot}3 \text{ cm}^2. \text{ Use } 45 \times 2 \text{ cm plates.}$$

3·13·1. Shear Stress

$$q_s = \frac{S}{dt_w} = \frac{56{,}200}{60 \times 1{\cdot}05} = 890 \text{ kg/cm}^2 < 945 \text{ kg/cm}^2$$

3·14. Trussed Girder

Span $= 54$ m

Depth of main girder $\not< \frac{1}{10}$th of effective span

∴ Height of truss $= 1/6$ to $1/7$ of the span $= 9$ m to $7{\cdot}7$ m

Use $h = 8$ m

Spacing of the main girders $= 8$ m $> 1/20$ of span

Depth of girder $\not>$ 3 times the spacing of the main girders $\not>$ $3 \times 8 = 24$ m

N—type of truss with cross girders at 6 m shall be used.

3·14·1. Loads

Dead loads

Due to wearing coat, deck slab, stringers and cross girders acting at each support $= 19200$ kg

Self weight of truss $= w = 15\,L + 550$ kg/m $= 15 \times 54 + 550$
$= 1360$ kg/m

This can be assumed to act at each model point

Force at each work $= 1360 \times 6 = 8160$ kg

Total load at each joint $= 19{\cdot}2 + 8{\cdot}16 = 27{\cdot}36$ t Say 28 t

Live loads. Class—A loading

1·35 t		3·5 t		5·7		5·7		3·4								1·35 t
↓	1.1 m	↓	3·2 m	↓	1·2 m	↓	4·3 m	↓	3 m	↓	3 m	↓	3 m	↓	20m	↓
(1)		(2)		(3)		(4)		(5)		(6)		(7)		(8)		
(9)		(10)		(11)		(12)		(13)		(14)		(15)		(16)		

Maximum effect is produced due to moving load when two trains side by side cross or pass on the bridge for the arrangement shown.

For any given axle load W the load transferred to the truss

$$=\frac{W}{8}\ (7+5{\cdot}2+3{\cdot}78+1{\cdot}98)=\frac{W}{8}\times 17{\cdot}96=2{\cdot}24\ w$$

3·14·2. Impact Factor

For spans greater than 45 m, impact factor=0.154

Live load, due to any axle ioad w on the truss, including impact $=1{\cdot}154\times 2{\cdot}24\ w=2{\cdot}59\ w$

3·14·3. Maximum Live Loads on the Truss

3·5 t		3·5 t		14·75 t		14·75		8·8		8·8		8·8		8·8		3·5
↓	1·1 m	↓	3·2 m	↓	1·2 m	↓	4·3m	↓	3 m	↓	3 m	↓	3 m	↓	20 m	↓
(1)		(2)		(3)		(4)		(5)		(6)		(7)		(8)		
(9)		(10)		(11)		(12)		(13)		(14)		(15)		(16)		

3·15. Forces In Chord Members

Centre of moment U_1,

$$IL \text{ ordinate}=\frac{\text{BM about } U_1}{\text{ht. of truss}}=\frac{6\times 48}{54\times 8}=0\ 66$$

Force in $L_1\,L_2$ due to live load including impact.

The force is maximum when the average load on the left of L_1 and average load on the right of L_1 are equal.

3.15.1. Force in $L_1\,L_2$

$$\frac{W_1}{x_1}-\frac{W_2}{x_2} \text{ should change sign}$$

Try load (3) at the section

$$\frac{W_1}{x^1}=\frac{3{\cdot}5+3{\cdot}5+14{\cdot}75}{6}=\frac{21{\cdot}75}{6}=3{\cdot}62$$

$$\frac{W_2}{x_2}=\frac{14{\cdot}75+8{\cdot}8\times 4+3{\cdot}5\times 2+14{\cdot}75\times 2+8{\cdot}8\times 2}{48}$$

$$=\frac{104{\cdot}05}{48}=2{\cdot}175$$

$$\frac{W_1}{x_1}=\frac{W_2}{x_2}=3\ 620-2{\cdot}175\ +\text{ve}$$

Shifting the load (3) on to the right

$$\frac{7}{6}-\frac{118{\cdot}8}{48}=1{\cdot}168-2{\cdot}48 \qquad -\text{ve}$$

Therefore the force in member $L_1\ L_2$ will be maximum when load (3) is at the section

Force in $L_1\ L_2$

$$=0{\cdot}66\left(3{\cdot}5\times\frac{1{\cdot}7}{6}+3{\cdot}5\times\frac{2{\cdot}8}{6}+14{\cdot}75+14{\cdot}75\times\frac{46{\cdot}8}{48}\right.$$

$$+8{\cdot}8\times\frac{42{\cdot}5}{48}+8{\cdot}8\times\frac{39{\cdot}5}{48}+8{\cdot}8\times\frac{36{\cdot}5}{48}+8{\cdot}8\times\frac{33{\cdot}5}{48}$$

$$+3{\cdot}5\times\frac{13{\cdot}5}{48}+3{\cdot}5\times\frac{12{\cdot}4}{48}+14{\cdot}75\times\frac{9{\cdot}2}{48}+14{\cdot}75\times\frac{8}{48}$$

$$\left.+8{\cdot}8\times\frac{3{\cdot}7}{48}+8{\cdot}8\times\frac{0{\cdot}7}{48}\right)$$

$$=0{\cdot}66\,(0{\cdot}89+1{\cdot}635+14{\cdot}75+14{\cdot}4+7{\cdot}8+7{\cdot}25+6{\cdot}7+6{\cdot}15$$

$$+0{\cdot}985+{\cdot}905+2{\cdot}83+2{\cdot}45+0{\cdot}68+0{\cdot}129)$$

$$=0{\cdot}66\times67{\cdot}654=4{\cdot}46 \text{ t (tension)}$$

Force in $L_1\ L_2$ due to dead loads

$$=\frac{28\times0{\cdot}66}{48}\,(48+42+36+30+24+18+12+6)$$

$$=\frac{28\times0{\cdot}66}{48}\times216=85{\cdot}6 \text{ t (tension)}$$

3·15·2. Force In $L_0\ L_1$

Force in $L_0\ L_1$ is same as in $L_1\ L_2$. Centre of moment is U_2

$$\text{IL Ordinate}=\frac{x(L-x)}{lh}=\frac{12\times42}{54\times8}=7/6$$

3·15·3. Force in $L_2\ L_3$

Force due to live load in member $L_2\ L_3$

For force to be maximum in $L_2\ L_3$,

$\dfrac{w_1}{x_1}-\dfrac{w_2}{x_2}$ shhould change sign

Try load (4) at the section

$$\frac{36{\cdot}5}{12}-\frac{72{\cdot}10}{42}=3{\cdot}04-1{\cdot}715 \qquad +\text{ve}$$

$$\frac{21{\cdot}75}{12}-\frac{86{\cdot}85}{42}=1{\cdot}81-2{\cdot}96 \qquad -\text{ve}$$

Therefore the force in member $L_2\ L_3$ to be maximum load (4) ould be placed at the section.

Force in $L_2\ L_3$

$$=\frac{7}{6}\left(3{\cdot}5\times\frac{6{\cdot}5}{12}+3{\cdot}5\times\frac{7{\cdot}6}{12}+14{\cdot}75\times\frac{10{\cdot}8}{12}+14{\cdot}75\right.$$

$$+8{\cdot}8\times\frac{37{\cdot}7}{42}+8{\cdot}8\times\frac{34{\cdot}7}{42}+8{\cdot}8\times\frac{31{\cdot}7}{42}+8{\cdot}8\times\frac{28{\cdot}7}{42}$$

$$\left.+3{\cdot}5\times\frac{8{\cdot}7}{42}+3{\cdot}5\times\frac{7{\cdot}6}{42}+14{\cdot}75\times\frac{4{\cdot}4}{42}+14{\cdot}75\times\frac{3{\cdot}2}{42}\right)$$

$=7/6\,(1{\cdot}90+2{\cdot}22+13{\cdot}30+14{\cdot}75+7{\cdot}9+7{\cdot}25+6{\cdot}65+6{\cdot}05$
$+0{\cdot}725+0{\cdot}634+1{\cdot}55+1{\cdot}15)$
$=7/6\times 64{\cdot}079=74{\cdot}8$ t (tension)

Force due to dead load in $L_2\,L_3$

$$=\frac{28\times 7}{6}\left(\frac{6}{12}+1+\frac{36}{42}+\frac{30}{42}+\frac{24}{42}+\frac{18}{42}+\frac{12}{42}+\frac{6}{42}\right)$$

$$=\frac{28\times 7}{6}\left(1{\cdot}5+\frac{126}{42}\right)=\frac{28\times 7}{6}\times 4{\cdot}5=147 \text{ t (tension)}$$

3·15·4. Force in $L_3\,L_4$

Centre of moment U_3

Influence live ordinate

$$=\frac{\text{BM about } U_3}{ht \text{ of truss}}=\frac{x(L-x)}{l\times h}=\frac{18\times 36}{54\times 8}=1{\cdot}5$$

Force due to live load including impact in member $L_3\,L_4$

For force to be maximum $L_3\,L_4$, the condition is

$\frac{W_1}{x_1}-\frac{W_2}{x_2}$ should change sign

Try load (5) at the section

$\frac{36{\cdot}50}{18}-\frac{71{\cdot}7}{36}=2{\cdot}05-1{\cdot}96$ +ve

$\frac{45{\cdot}3}{18}-\frac{62{\cdot}9}{36}=2{\cdot}52-1{\cdot}71$ +ve

Try load (4) $\frac{36{\cdot}50}{18}-\frac{42{\cdot}2}{36}=2{\cdot}05-1{\cdot}175$ +ve

$\frac{21{\cdot}75}{18}-\frac{56{\cdot}95}{36}=1{\cdot}21-1{\cdot}58$ −ve

For Force to be maximum in $L_3\,L_4$ load (4) should be placed at the secticn. Force in $L_3\,L_4$

$$=1{\cdot}5\left(3{\cdot}5\times\frac{12{\cdot}5}{16}+3{\cdot}5\times\frac{13{\cdot}6}{18}+14{\cdot}75\times\frac{16{\cdot}8}{18}\right.$$
$$+14{\cdot}75+8{\cdot}8\times\frac{31{\cdot}7}{36}+8{\cdot}8\times\frac{28{\cdot}7}{36}+8{\cdot}8\times\frac{25{\cdot}7}{36}$$
$$\left.+8.8\times\frac{23{\cdot}7}{36}+3{\cdot}5\times\frac{2{\cdot}7}{6}+3{\cdot}5\times\frac{1{\cdot}6}{36}\right)$$

$=1{\cdot}5(2{\cdot}43+2{\cdot}64+13{\cdot}78+14{\cdot}75+7{\cdot}75+7{\cdot}02+6{\cdot}28+5.55+{\cdot}262$
$+{\cdot}055)=1{\cdot}5\times 60{\cdot}517=90{\cdot}8$ t (tensile)

Force due to dead load in $L_3\,L_4$

$$=28\times 1{\cdot}5\left(\frac{6}{18}+\frac{12}{18}+1+\frac{30}{36}+\frac{24}{36}+\frac{18}{36}+\frac{12}{36}+\frac{6}{36}\right)$$

$=28\times 1{\cdot}5\,(2+2{\cdot}5)=28\times 1{\cdot}5\times 4{\cdot}5=189$ t (tension)

3·15·5. Force in $L_4\,L_5$

Centre of momemt U_4

Influence live ordinate$=\dfrac{\text{BM about } U_4}{\text{height of the truss}}=\dfrac{20\times 30}{54\times 8}=\dfrac{5}{3}$

Force due to live load including impact in member $L_4\,L_5$

For force to be maximum in $L_3\,L_4$, the condition is

$\dfrac{W_1}{x_1}-\dfrac{W_2}{x_2}$ should change sign.

Try load (4) $\dfrac{36{\cdot}50}{24}-\dfrac{55{\cdot}2}{30}=1{\cdot}52-1{\cdot}17\quad +\text{ve}$

$\dfrac{21{\cdot}75}{24}-\dfrac{49{\cdot}95}{30}=0{\cdot}91\ -1{\cdot}66\quad -\text{ve}$

Force in $L_4\,L_5$

$=\dfrac{5}{3}\left(3{\cdot}5\times\dfrac{18{\cdot}5}{24}+3{\cdot}5\times\dfrac{19{\cdot}6}{24}+14{\cdot}75\times\dfrac{22{\cdot}8}{24}+14{\cdot}75\right.$

$\left.+8{\cdot}8\times\dfrac{25{\cdot}7}{30}+8{\cdot}8\times\dfrac{22{\cdot}7}{30}+8{\cdot}8\times\dfrac{19{\cdot}7}{30}+8{\cdot}8\times\dfrac{16{\cdot}7}{30}\right)$

$=5/3\ (2{\cdot}79+2{\cdot}86+14+14{\cdot}75+7{\cdot}55+6{\cdot}66+5{\cdot}78+4{\cdot}9)$

$=5/3+59{\cdot}29=98{\cdot}9$ t (tension)

Force due to dead load in $L_4\,L_5$

$=28\times\dfrac{5}{3}\left(\dfrac{6}{24}+\dfrac{12}{24}+\dfrac{18}{24}+1+\dfrac{24}{30}+\dfrac{18}{30}+\dfrac{12}{30}+\dfrac{6}{30}\right)$

$=28\times\dfrac{5}{3}\left({\cdot}25+0{\cdot}5+0{\cdot}75+1+\dfrac{60}{30}\right)=28\times\dfrac{5}{3}\ (4{\cdot}5)$

$=210$ t (tension)

3·16. Top Chord Members $U_1\,U_2$

Force in $U_1\,U_2$ (centre of moment is L_2)

I.L Ordinate$=\dfrac{\text{BM about } L_2}{\text{height of truss}}=\dfrac{12\times 42}{8\times 54}=\dfrac{7}{6}$

IL is same as member $L_2\,L_3$

Force due to live loads$=74{\cdot}8$ t (comp)

Force due to dead loads$=147$ t (comp)

3·16·1. Force in $U_2\,U_3$

Centre of moment L_3

IL ordinate$=\dfrac{18\times 36}{54\times 8}=1{\cdot}5$

IL is same as the member $L_3\,L_4$

Force in $U_2\,U_3$ due to live loads$=90{\cdot}8$ t (comp)

Force in $U_2\,U_3$ due to dead loads$=189$ t (comp)

3·16·2. Force in $U_3\,U_4$, $U_4\,U_5$

Centre of moment L_4

IL ordinate$=\dfrac{24\times 30}{54\times 8}=\dfrac{5}{3}$

IL is same as the member $L_4\,L_5$

Force due to live loads=98·9 t (comp)
Force due to dead loads=210 t (comp)

3·17. Diagonal Members : Force in $L_0\ U_1$

IL for force in $L_0\ L_1$

Shear in the panel × sec θ

$$\text{Sec } \theta = \frac{10}{8} = 1{\cdot}25$$

$$\text{IL ordinate} = 1{\cdot}25 \times \frac{48}{54} = 1{\cdot}11$$

For force to be maximum in $L_0\ U_1$

$\frac{W}{L} - \frac{w}{l}$ should change sign

Where W=total load on the span, L=span, w=load on the left of section, l=length of span to the left of section.

Try load (3) at the section $\frac{W}{L} - \frac{w}{l} = \frac{135{\cdot}8}{54} - \frac{21{\cdot}75}{6}$

$=2{\cdot}52-3{\cdot}62= -\text{ve}$; $\quad \frac{135{\cdot}8}{54} - \frac{7}{6} = 2{\cdot}52-1{\cdot}167=+\text{ve}$

For force in $L_0\ U_1$ is maximum when load (3) is at L_1

Force due to live load in $L_0\ U_1$

$$=1{\cdot}11\Big(3{\cdot}5\times\frac{1{\cdot}7}{6}+3{\cdot}5\times\frac{2{\cdot}8}{6}+14{\cdot}75+14{\cdot}75$$
$$\times\frac{46{\cdot}8}{48}+8{\cdot}8\times\frac{42{\cdot}5}{48}+8{\cdot}8\times\frac{39{\cdot}5}{48}+8{\cdot}8\times\frac{36{\cdot}5}{48}$$
$$+8{\cdot}8\times\frac{33{\cdot}5}{48}+3{\cdot}5\times\frac{13{\cdot}5}{48}+3{\cdot}5\frac{12{\cdot}4}{48}$$
$$+14{\cdot}75\times\frac{9{\cdot}2}{48}+14{\cdot}75\times\frac{8}{48}+8{\cdot}8\times\frac{3{\cdot}7}{48}+8{\cdot}8\times\frac{0{\cdot}7}{48}\Big)$$

$=1{\cdot}11\ (0{\cdot}99+1{\cdot}635+14{\cdot}75+14{\cdot}40+7{\cdot}8+7{\cdot}25$
$\quad +6{\cdot}70+6{\cdot}15+0{\cdot}985+{\cdot}905+2{\cdot}83+2{\cdot}45+0{\cdot}68+0{\cdot}129)$

$=1{\cdot}11\times67{\cdot}654=75{\cdot}4$ t (comp)

Force due to dead loads

$$=28\times1{\cdot}11\Big(1+\frac{42}{48}+\frac{36}{48}+\frac{30}{48}+\frac{24}{48}+\frac{18}{48}+\frac{12}{48}+\frac{6}{48}\Big)$$

$=28\times1{\cdot}11\times4{\cdot}5=139$ t (comp).

3·17·1. Force in $U_1\ L_2$

Force in $U_1\ U_2$=shear in the panel $L_1\ L_2$ × sec θ

For tensile force to be maximum

Try load (4) at L_2. $\frac{W}{L} - \frac{W}{l}$ should change sign

$$\frac{108{\cdot}2}{54} - \frac{21{\cdot}75}{12} = 2{\cdot}01-1{\cdot}81 \quad +ve$$

$\frac{108\cdot2}{54}-\frac{36\cdot5}{12}=2\cdot01-3\cdot04 \quad -ve$

Tensile force in the member due to LL

$=0\cdot98\Big[3\cdot5\times\frac{\cdot89}{5\cdot29}+14\cdot75\times\frac{4\cdot09}{5\cdot29}+14\cdot75$

$+8\cdot8\times\frac{37\cdot7}{42}+8\cdot8\times\frac{34\cdot7}{42}+8\cdot8\times\frac{31\cdot7}{42}$

$+8\cdot8\times\frac{28\cdot7}{42}+3\cdot5\times\frac{8\cdot7}{42}+3\cdot5\times\frac{7\cdot1}{42}+\frac{14\cdot75\times3\cdot9}{42}$

$+14\cdot75\times\frac{2\cdot7}{42}\Big]-3\cdot5\times\frac{\cdot21}{\cdot71}\times0\cdot14$

$+1\cdot83+\cdot59+1\cdot37+\cdot95)-3\cdot5\times\cdot042$

$=\cdot98[\cdot590+11\cdot40+14\cdot75+7\cdot9+7\cdot3+6+1\cdot83+\cdot59+1\cdot37$

$+\cdot95)-3\cdot5\times\frac{\cdot21}{\cdot71}\times0\cdot14$

$=\cdot98\times52\cdot68-\cdot145=51\cdot5$ t (tensile) $-\cdot145$ t$=51\cdot355$ t

Maximum compressive force will occur when only one vehicle is on the span. Such that the either load (6) or (7) at the section

Compressive force in the member$=0\cdot14\ (8\cdot8+8\cdot8\times\frac{3}{6})$

$=0\cdot14\ (8\cdot8\times4\cdot4)=0\cdot14\times13\cdot2=1\cdot85$ t (comp)

Force due to dead load

Force in $U_1L_2=+0\cdot14\times28-\cdot98\times28\left(1+\frac{36}{42}+\frac{30}{42}\right.$

$\left.+\frac{24}{42}+\frac{18}{42}+\frac{12}{42}+\frac{6}{42}\right)$

$=0\cdot14\times28-0\cdot98\times28\times4=3\cdot92-110=-106\cdot08$ t (tensile)

3·17·2. Force in U_2 L_3

Force in U_2L_3=shear in the panel $L_2L_3\times\sec\theta$

For shear to be maximum in the panel

$\frac{W}{L}-\frac{W}{l}$ should change sign.

Influence live ordinate at $L_2=\frac{12}{54}\sec\theta=\frac{12}{54}\times1\cdot25=27\cdot8=28$

IL ordinate at $L_3=\frac{36}{54}\sec\theta=\frac{36}{54}\times1\cdot25=\cdot835=\cdot84$

Try load (4) at L_3 for maximum tensile force to occur in U_2L_3

$\frac{78\cdot2}{54}-\frac{21\cdot75}{18}=1\cdot45-1\cdot21=+ve$

$\frac{78\cdot2}{54}-\frac{36\cdot5}{18}=1\cdot45-2\cdot02=-ve$

Maximum tensile force is produced when load (4) is at L_3

Tensile force in $U_2 L_3$ due to live load

$$=0{\cdot}84\left(3{\cdot}5\times\frac{0{\cdot}1}{4{\cdot}5}\times14{\cdot}75+14{\cdot}75+8{\cdot}8\times\frac{31{\cdot}7}{36}\right.$$

$$+8{\cdot}8\times\frac{28{\cdot}7}{36}+8{\cdot}8\times\frac{25{\cdot}7}{36}+8{\cdot}8\times\frac{22{\cdot}7}{36}+3{\cdot}5\times\frac{2{\cdot}7}{36}$$

$$\left.+3{\cdot}5\times\frac{1{\cdot}6}{36}\right)-\frac{1}{1{\cdot}5}\times0{\cdot}28\times3{\cdot}5$$

$$=0{\cdot}84\,({\cdot}0778+10{\cdot}8+14{\cdot}75+7{\cdot}75+7{\cdot}02+6{\cdot}28$$
$$+5{\cdot}55+0{\cdot}262+{\cdot}155)-{\cdot}65$$

$$=0{\cdot}84\times52{\cdot}2278-0{\cdot}65=43{\cdot}2\text{ t (tensile)}$$

Maximum compressive force is produced when load (8) is at L_2

Compressiue force in $U_2 L_3$ due to live load

$$=0{\cdot}28\left(8{\cdot}8+8{\cdot}8\times\frac{9}{12}+8{\cdot}8\times\frac{6}{12}+8{\cdot}8\times\frac{3}{12}\right)$$

$$=0{\cdot}28\times8{\cdot}8(1+0{\cdot}75+0{\cdot}5+0{\cdot}25)=0{\cdot}28\times8{\cdot}8\times2{\cdot}5=6{\cdot}35\text{ t (comp)}$$

Force due to dead loads

Force in $U_2 L_3$

$$=0{\cdot}28\times28\left(\frac{6}{12}+1\right)-28\times{\cdot}84\left(1+\frac{30}{36}+\frac{24}{36}\right.$$
$$\left.+\frac{18}{36}+\frac{12}{36}+\frac{6}{36}\right)$$

$$=0{\cdot}28\times28\times1{\cdot}5-28\times{\cdot}84\times3{\cdot}5=11{\cdot}8-82{\cdot}5=70{\cdot}7\text{ t (tension).}$$

3·17·3. Force in $U_3 L_4$

Force in $U_3 L_4$=shear in the panel $L_3 L_4\times\sec\theta$

$$\text{I.L. ordinate at } L_3=\frac{18}{54}\times\sec\theta=1{\cdot}25\times\frac{18}{54}=0{\cdot}425$$

$$\text{I.L. ordinate at } L_4=\frac{30}{54}\times\sec\theta=1{\cdot}25\times\frac{30}{54}=0{\cdot}695$$

Try load (4) at L_4

$\frac{W}{L}-\frac{W}{l}$ should change sign

$$\frac{78{\cdot}2}{54}-\frac{21{\cdot}75}{24}=1{\cdot}45-{\cdot}905\quad+ve$$

$$\frac{78{\cdot}2}{54}-\frac{36{\cdot}5}{24}=1{\cdot}45-1{\cdot}52\quad-ve$$

Maximum tensile force in $U_3 L_4$ occurs when load (4) in at L_4

Tensile force in $U_3 L_4$

$$=0{\cdot}695\left(\frac{2{\cdot}54}{3{\cdot}74}\times14{\cdot}75+14{\cdot}75+8{\cdot}8\times\frac{25{\cdot}7}{30}+8{\cdot}8\times\frac{22{\cdot}7}{30}\right.$$

$$\left.+8{\cdot}8\times\frac{18{\cdot}7}{30}+8{\cdot}8\times\frac{15{\cdot}7}{30}\right)-{\cdot}425\left(3{\cdot}5\times\frac{1{\cdot}76}{2{\cdot}26}+3{\cdot}5\times\frac{{\cdot}66}{2{\cdot}26}\right)$$

$=0{\cdot}695(10{\cdot}0+14{\cdot}75+7{\cdot}55+6{\cdot}65+5{\cdot}48+4{\cdot}6)$
$\quad -{\cdot}425(2{\cdot}73+1{\cdot}02)$
$=0{\cdot}695\times 49{\cdot}03-{\cdot}425\times 3{\cdot}75=33{\cdot}8-1{\cdot}6=32{\cdot}2$ t (Tensile).

Maximum compressive force is produced when load (8) is at L_3

Compressive force in $U_3\ L_4$ due to live load

$$={\cdot}42\left(8{\cdot}8+8{\cdot}8\times\frac{15}{18}+8{\cdot}8\times\frac{12}{18}+8{\cdot}8\times\frac{9}{18}\right.$$
$$\left.+14{\cdot}75\times\frac{4{\cdot}7}{18}+14{\cdot}75\times\frac{3{\cdot}5}{18}+3{\cdot}5\times\frac{3}{18}\right)$$

$={\cdot}425(8{\cdot}8+7{\cdot}35+5{\cdot}86+4{\cdot}4+3{\cdot}85+2{\cdot}87+{\cdot}582)$
$=0{\cdot}425\times 33{\cdot}312=14{\cdot}1$ t (comp)

Force due to dead loads

$$=0{\cdot}425\times 28\left(\frac{6}{18}+\frac{12}{18}+1\right)-0{\cdot}695\times 28$$
$$\left(1+\frac{24}{30}+\frac{18}{30}+\frac{12}{30}+\frac{6}{30}\right)$$

$=0{\cdot}425\times 28\times 2-0{\cdot}695\times 28\times 3=23{\cdot}6-58{\cdot}5=34{\cdot}9$ t (tensile)

3·17·4. Force in $U_4\ L_5$

Force in $U_3\ L_5$=shear in the panel $L_4\ L_5\times\sec\theta$

I.L ordinate at $L_4=\frac{24}{54}\times 1{\cdot}25={\cdot}56$

I.L ordinate at $L_5=\frac{24}{54}\times 1{\cdot}25=0{\cdot}56$

Try load (4) at L_5, $\frac{W}{L}-\frac{w}{l}$ should change sign

$\frac{80{\cdot}5}{54}-\frac{30{\cdot}55}{30}=1{\cdot}49-1{\cdot}01=$ +ve

$\frac{80{\cdot}5}{54}-\frac{45{\cdot}3}{30}=1{\cdot}49-1{\cdot}51=$ −ve

Max tensile force in $U_4\ L_5$ occurs when load (4) is at L_5

$$\text{Tensile force}=0{\cdot}56\left(14{\cdot}75\times\frac{1{\cdot}8}{3}+14{\cdot}75+8{\cdot}8\times\frac{19{\cdot}7}{24}\right.$$
$$\left.+8{\cdot}8\times\frac{16{\cdot}7}{24}+8{\cdot}8\times\frac{15{\cdot}9}{24}+8{\cdot}8\times\frac{10{\cdot}7}{24}\right)$$
$$-{\cdot}56\left(\frac{3{\cdot}5\times 1{\cdot}4}{3}+\frac{3{\cdot}4\times 2{\cdot}5}{3}\right)$$

$=0{\cdot}56(8{\cdot}85+14{\cdot}75+7{\cdot}25+6{\cdot}15+5{\cdot}02+1{\cdot}95$
$\quad -{\cdot}56(1{\cdot}64+2{\cdot}82)$
$=0{\cdot}56\times 45{\cdot}97-{\cdot}56\times 4{\cdot}46=25{\cdot}6-2{\cdot}5=23{\cdot}1$ (Tensile)

For maximum Compressive force in $U_4\ L_5$

Maximum compressive force is produced when load (8) in at L_4

Compressive force in $U_4 L_5$ due to live load

$$=0{\cdot}56\Big(8\ 8+8{\cdot}8\times\frac{21}{24}+8{\cdot}8\times\frac{15}{24}+8{\cdot}8\times\frac{15}{24}$$

$$-14{\cdot}75\times\frac{11{\cdot}7}{24}+14{\cdot}75\times\frac{10{\cdot}5}{24}+3{\cdot}5\times\frac{7{\cdot}3}{24}+3{\cdot}5\times\frac{6{\cdot}2}{24}\Big)$$

$$=0{\cdot}56(8{\cdot}8+7{\cdot}7+6{\cdot}6+5{\cdot}5+7{\cdot}2+6{\cdot}45+1{\cdot}06+{\cdot}905)$$

$$=0{\cdot}56\times 44{\cdot}215=24{\cdot}7 \text{ t (comp)}$$

Force due to dead loads

$$=28\times 0{\cdot}56\Big(\frac{6}{24}+\frac{12}{24}+\frac{18}{24}+1\Big)-28\times{\cdot}56\times$$

$$\Big(1+\frac{18}{24}+\frac{12}{24}+\frac{6}{24}\Big)=0$$

3·18. Vertical members force in $U_1 L_1$

Force in $U_1 L_1$=Shear in the panel L_1L_2

Shear in a panel is maximum when $\frac{W}{L}-\frac{w}{l}$ change sign

Try load (3) at L_1, $\frac{125{\cdot}80}{54}-\frac{21{\cdot}75}{6}=2{\cdot}34-3\ 62=$ −ve

$\frac{125{\cdot}80}{54}-\frac{7}{6}=2{\cdot}34-1{\cdot}17=$ +ve

Tensile force in $U_1 L_1$ is maximum when load (3) is at L_1

$$\text{Force in } U_1 L_1=\frac{1{\cdot}7}{6}\times 3{\cdot}5+3{\cdot}5\times\frac{2{\cdot}8}{6}+14{\cdot}75\times\frac{6}{6}$$

$$+14{\cdot}75\times\frac{4\ 8}{6}+8{\cdot}8\times\frac{0{\cdot}5}{6}$$

$$={\cdot}99+1{\cdot}63+14{\cdot}75+11{\cdot}81+{\cdot}73=29{\cdot}90 \text{ t (tensile)}$$

Force due to dead loads

Force in $U_1 L=28\times 1=28$ t (Tensile)

3·18·1. Force in $U_2 L_2$

Force in Member $U_2 L_2$=shear in panel $L_2 L_3$

$I.L$ ordinate at $L_2=\frac{12}{54}=0{\cdot}222$

$I.L$ ordinate at $L_3=\frac{36}{54}=0{\cdot}666$

For shear force to be maximum in the panel

$\frac{W}{L}-\frac{w}{l}$ should change sign

Try load (4) at L_2, $\frac{108{\cdot}2}{54}-\frac{21\ 75}{12}=2{\cdot}01-1{\cdot}81=$ +ve

$\frac{108{\cdot}2}{54}-\frac{36{\cdot}5}{12}=2{\cdot}01-3{\cdot}04=$ −ve

Maximum tensile force occurs in U_2 L_2, when load (8) is at L_2 and only one vehicle is on the span.

Force in $U_2 L_2 = 0{\cdot}222 \times 8{\cdot}8\left(1+\frac{9}{12}+\frac{6}{12}+\frac{3}{12}\right)$

$= 0{\cdot}222 \times 8{\cdot}8 \times 2{\cdot}5 = 4{\cdot}66$ t (tensile)

Try load (4) at L_3 for maximum compressive force in $U_2 L_2$

$$\frac{78{\cdot}2}{54}-\frac{21{\cdot}75}{18}=1{\cdot}45-1{\cdot}21= \quad +\text{ve}$$

$$\frac{78{\cdot}2}{54}-\frac{36{\cdot}5}{18}=1{\cdot}45-2{\cdot}02= \quad -\text{ve}$$

Compressive force is maximum in $U_2 L_2$ when load (4) is at L_3

$$= 0{\cdot}666\Big[\Big(3{\cdot}5\times\frac{0{\cdot}1}{4.6}+14{\cdot}75\times\frac{3{\cdot}3}{4{\cdot}5}+14{\cdot}75$$

$$+8{\cdot}8\times\frac{31{\cdot}7}{36}+8{\cdot}8\times\frac{28{\cdot}7}{36}+8{\cdot}8\times\frac{25{\cdot}7}{36}$$

$$+8{\cdot}8\times\frac{22{\cdot}7}{36}+3{\cdot}5\times\frac{2{\cdot}7}{36}+3{\cdot}5\times\frac{1{\cdot}6}{36}\Big)\Big]$$

$$-{\cdot}222\times\frac{1}{1{\cdot}5}\times 3{\cdot}5$$

$= 0{\cdot}666[({\cdot}0778+10{\cdot}8+14{\cdot}75+7{\cdot}75+7{\cdot}02$

$+6{\cdot}28+5{\cdot}55+0{\cdot}262+0{\cdot}155)]-{\cdot}518$

$= 0{\cdot}666\times 52{\cdot}2278-{\cdot}518=34{\cdot}8-{\cdot}518=34{\cdot}282$ t (comp)

Dead load forces

Forces in $U_2 L_2 = -0{\cdot}222\times 28\left(\frac{6}{12}+1\right)+28\times 0{\cdot}666$

$$\left(1+\frac{30}{36}+\frac{24}{36}+\frac{18}{36}+\frac{12}{36}+\frac{6}{36}\right)$$

$= 0{\cdot}222\times 28\times 1{\cdot}5+28\times 0{\cdot}666\times 3{\cdot}5=-9{\cdot}34+65{\cdot}5$

$= 55{\cdot}16$ t (comp)

3·18·2. Force in $U_3 L_3$

Force in member $U_3 L_3$ = Shear in the panel $U_3 L_4$

$I L$ ordinate at $L_3 = \frac{18}{54} = 0{\cdot}333$

$I L$ ordinate at $L_4 = \frac{30}{54} = 0{\cdot}555$

Try load (4) at $L_4 = \frac{W}{L} - \frac{w}{l}$ should change sign

$$\frac{78{\cdot}2}{54}-\frac{21{\cdot}75}{24}=1.45-0{\cdot}905= \quad +\text{ve}$$

$$\frac{78{\cdot}2}{54}-\frac{36{\cdot}5}{24}=1{\cdot}45-1{\cdot}52= \quad -\text{ve}$$

The maximum comp.force in U_3L_4 occurs when load (4) is at L_4

Compressive force in $U_3\ L_3$.

$$= 0{\cdot}555\left(\frac{2{\cdot}54}{3{\cdot}74}\times 14{\cdot}75+14{\cdot}75+8{\cdot}8\times\frac{25{\cdot}7}{30}\right.$$

$$\left.+8{\cdot}8\times\frac{22{\cdot}7}{30}+8{\cdot}8\times\frac{18{\cdot}7}{30}+8{\cdot}8\ \frac{15{\cdot}7}{30}\right)$$

$$-{\cdot}333\left(3{\cdot}5\times\frac{1{\cdot}76}{2{\cdot}26}+3{\cdot}5\times\frac{{\cdot}66}{2{\cdot}26}\right)$$

$$=0{\cdot}555(10\times 14{\cdot}75+7{\cdot}55+6{\cdot}65+5{\cdot}48+4{\cdot}6)-{\cdot}333(3{\cdot}75)$$

$$=0{\cdot}555\times 49{\cdot}03-1{\cdot}25=27{\cdot}4-1{\cdot}25=26{\cdot}15 \text{ t (comp)}$$

For maximum tensile force in U_3L_3 occurs when load (8) is at L_3

Tensile force in $U_3\ L_3$ due to live load

$$=0{\cdot}333\left((8{\cdot}8+8{\cdot}8\times\frac{15}{18}+8{\cdot}8\times\frac{12}{18}+8{\cdot}8\times\frac{9}{18}\right.$$

$$\left.+14{\cdot}75\times\frac{4{\cdot}7}{18}+14{\cdot}75\times\frac{3{\cdot}5}{18}+3{\cdot}5\times\frac{{\cdot}3}{18}\right)$$

$$=0{\cdot}333\times(8{\cdot}8+7{\cdot}35+5{\cdot}86+4{\cdot}4+3{\cdot}85+2{\cdot}87+0{\cdot}582)$$

$$=0{\cdot}333\times 33{\cdot}312=11{\cdot}1 \text{ t (tensile)}$$

Force due to dead load

$$=-0{\cdot}333\times 28\left(\frac{6}{18}+\frac{12}{18}+1\right)+0{\cdot}555\times 28$$

$$\left(1+\frac{24}{30}+\frac{18}{30}+\frac{12}{30}+\frac{6}{30}\right)$$

$$=-0{\cdot}333\times 28\times 2+0{\cdot}555\times 28\times 3=-18{\cdot}65+46{\cdot}6=27{\cdot}95 \text{ t}$$

(comp)

3·18·3. Force in $U_4\ L_4$

Force $-U_4\ L_4$=shear in the panel $L_4\ L_5$

$$I\,L \text{ ordinate at } L_4=\frac{24}{54}={\cdot}444$$

$$I\,L \text{ ordinate at } L_5=\frac{24}{54}={\cdot}444$$

Try load (4) at L_5

$\frac{W}{L}-\frac{w}{l}$ should change sign

$$\frac{80{\cdot}5}{54}-\frac{30{\cdot}55}{30}=1{\cdot}49-1{\cdot}01=\ +\text{ve}$$

$$\frac{80{\cdot}5}{54}-\frac{45{\cdot}3}{30}=1{\cdot}49-1{\cdot}51=\ -\text{ve}$$

Maximum compressive force in $U_4\ L_4$ occurs when load (4) is at L_5

$$\text{Compressive force in } U_4\ L_4=0{\cdot}444\left(14{\cdot}75\times\frac{1{\cdot}8}{3}+14{\cdot}75+8{\cdot}8\right.$$

$$\left.\times\frac{19{\cdot}7}{24}+8{\cdot}8\times\frac{16{\cdot}7}{24}+8{\cdot}8\times\frac{13{\cdot}7}{24}+8{\cdot}8\times\frac{10{\cdot}7}{24}\right)$$

$$-{\cdot}444\left(3{\cdot}5\times\frac{1{\cdot}4}{3}+3{\cdot}5\times\frac{2{\cdot}5}{3}\right)$$

$=\cdot444(8\cdot85+14.75+7\cdot25+6\cdot15+5\cdot02+3\cdot95)-0\cdot444(1\cdot64+282)$
$=0\cdot444\times45\cdot97-0\cdot444\times4\cdot46=20\cdot4-1\cdot98=18\cdot42$ t (comp)

For maximum tensile force in $U_4\,L_4$

Maximum tensile force is produced when load (8) is at L_4.

Tensile force in $U_4\,L_4$ due to live load

$$=0\cdot444\left(8\cdot8+8\cdot8\times\frac{21}{24}+8\cdot8\times\frac{18}{24}+8\cdot8\times\frac{15}{24}+14\cdot75\times\frac{11\cdot7}{24}+14\cdot75\times\frac{10\,5}{24}+3\cdot5\times\frac{7\cdot3}{24}+3\cdot5\times\frac{6\cdot2}{24}\right)$$

$=0\cdot444(8\cdot8+7\cdot7+6\cdot6+5\cdot5+7\cdot2+6\cdot45+1\cdot06+\cdot905)$
$=0\cdot444\times44\cdot215=19\cdot67$ (Tensile)

Force due to dead loads

$$=-28\times0\cdot444\left(\frac{6}{24}+\frac{12}{24}+\frac{18}{24}+1\right)+28\times\cdot444\left(1+\frac{18}{24}+\frac{12}{24}+\frac{6}{24}\right)=0$$

Member	*Live load*		*Dead loads*		*Combined load*		*Design loads*
	Comp	*Tension*	*Comp*	*Tension*	*Maximum* —*Ten* +*comp*	*Minimum* —*Ten* +*comp*	
Bottom chord members							
$L_0\,L_1$		44·6		85·6	−130·2	− 85·6	
$L_1\,L_2$		44·6		85·6	−130·2	− 85·6	
$L_2\,L_3$		74·8		147·0	−221·8	−147·0	
$L_3\,L_4$		80·8		189·0	−279·8	−189·0	
$L_4\,L_5$		98·9		210·0	−308·9	−210·0	
Top chord members							
$U_1\,U_2$	74·8		147		+221·8	+147	
$U_2\,U_3$	90·8		189		+279·8	+189	
$U_3\,U_4$	98·9		210		+308·0	+210	
$U_4\,U_5$	98·9		210		+308·9	+210	
Diagonal member							
$U_0\,L_1$	75·4		139		+214·4	+139·0	
$U_1\,L_2$	1·85	51·355		106·08	−157·435	−104·23	
$U_2\,L_3$	6·35	43·25		70·7	−119·95	− 64·35	
$U_3\,L_4$	14·1	32·2		24·9	− 57·1	− 10·8	
$U_4\,L_5$	24·7	23·1		0	− 23·1	+ 24·7	
Vertical membere							
$U_1\,L_1$		29·9		28·0	− 57·9	− 28·0	
$U_2\,L_2$	34·282	4·66	55·16		+ 89·442	+ 50·50	
$U_3\,L_3$	26·15	11·1	27·95		+ 54·10	+ 16·95	
$U_4\,L^4$	18·42	19·6	0		+ 18·42	− 19·6	

3·19. Design of truss members top boom members

Force in member $U_4 L_5$=308·9 t (comp)

Unsupported length of the member=6 m

Overall width of the member is $\nless$ 1/15 of the unsupported distance between the points of intersection of the lateral bracing or the length of compression members $\ngtr$ 45 times this width unless effective lateral stays are provided, $d/l \ngtr 9·0$.

3·19·1. Trial Section

Width of the section $\nless$ 1/15×600=40 cm

Use a section as shown

Approximate radius of gyration; $r_{xx}=0·4\ d=0·4\times70=28$ cm

$r_{yy}=0·34\times6=0·34\times60=20·4$ cm

effective length=0·85×600=510 cm

$$l/r_{min}=\frac{510}{20·4}=25$$

Allowable stress

for $l/r=20=P_{ac}=1360$ kg/cm²

$=40=P_{ac}=1300$ kg/cm²

for $l/r=25=P_{ac}=1345$ kg/cm²

$$\text{Area required for the section}=\frac{308900}{1345}=230 \text{ cm}^2$$

Using, 1—top flange plate 60 cm×1cm=60 cm²

2—side plates 70 cm×1 cm =140 cm²

4—angles ISA 7650 1 cm thick =4×11·52=46·08 cm²

Total area=246·08 cm²>230 cm²

Safe load carried by the member=246·08×1345=330 t

3·19·2. Lacing Specifications

Force in the member=308·9 t (compression)

Force on lacing =2½% of the force in the member.

As far as practicable, the lacing system shall not be varied throughout the length of the compression member.

Lacing bars shall be included at an angle of 50 to 70° to the axis of the member for single lacing and 40 to 50° in the case of double lacing.

The maximum spacing shall be such that l/r of the components of the compression member between consecutive connections of the lacing bars to one component is $\ngtr$ 50 or 0·7 times the maximum ratio of the slenderness of the member as a whole, whichever is lesser, where l=distance between the centres of connection. l/r ratio of the lacing bars $\ngtr$ 140.

Width of lacing bar shall not be less than double the edge distance.

3·19·3. Shear Force in Lacing Bars

$$\text{Force on lacing}=\frac{5}{200}\times308900=7720 \text{ kg}$$

Use single lacing system

Shear on lacing $=S=\frac{7720}{2}=3860$ kg

3·19·4. Spacing of Lacing Bars

$\frac{l}{r_{min}} \not> 50$

approximate r_{xx} for the component to which lacing is to be connected $=0·38d=0·38\times70=26·6$ cm

$r_{yy}=0·28b=·28\times8·5=2.38$ cm

$l \not> 2·38\times50=119$ cm

or $\frac{l}{r_{min}} \not> 0.7\times$ slenderness ratio of the entire section $\not> 0·7\times25$

$l=2·38\times0·7\times25=41·5$ cm

Use $=40$ cm

$\frac{40}{2·28}=16·8<50<0·7\times25=17·5$

3·19·5. Inclination of the Lacing

Tan $\theta=\frac{49·5}{20}=2·43$

$\theta=68°>40°<70°$

3·19·6. Force in Lacing Bar

Force in lacing $=\frac{S}{\sin\theta}=\frac{3860}{0·9272}=4180$ kg

3·19·7. Width of Lacing Bar

Using 22 mm diameter rivets for connecting the lacing, minimum width of lacing $=2\times$edge distance $=3d=3\times2·35=7·05$

Use 7·5 cm

3·19·8. Length of Lacing

$l=\frac{49·5}{\sin\theta}=\frac{49·5}{·9272}=53·4$ cm

$l/r \not> 140$

r_{min} required $=\frac{53·4}{140}=0·381$

r_{min} bar = flat $b\times t=\sqrt{\frac{bt^3}{12\times bt}}=\frac{t}{\sqrt{12}}$

$0·381\times\sqrt{12}=t$

The thickness of the lacing $=0·381\times3·46=1·32$ cm

Available size $=1·6$ cm

Use 7·5 cm $\times$ 1·6 cm flat as lacing bar

$r_{min}=\frac{1·6}{\sqrt{12}}=0·462$

$l/r_{min}=\frac{53·4}{0·462}=115<140$

P_{ac} allowable stress corresponding to
$l/r = 115 = 673$ kg/cm²

3·19·9. Safe Load on Lacing

Safe load on lacing $= 6{\cdot}5 \times 1{\cdot}6 \times 673 = 7000$ kg > 4180 kg

3·19·10. No. of Rivets for the Connection

Least strength of rivet in single shear

$$= \frac{\pi d^2}{4} f_s = \frac{\pi \times 2{\cdot}35^2}{4} \times 1020 = 4425 \text{ kg}$$

Number of rivets required $= \dfrac{4180}{4425} < 1$

Use one 22 mm diameter rivet

3·19·11. Tensile Stress in Lacing

$$\text{Tensile stress} = \frac{4180}{\text{Net area}} = \frac{4180}{(7{\cdot}5 - 2{\cdot}35) \times 16} = 630 \text{ kg/cm}^2$$
$$< 1510 \text{ kg/cm}^2$$

3·20. Bottom Boom Member

Force in member $L_4 L_5 = 308{\cdot}9$ t (tension)

Net area required $= \dfrac{308900}{1410} = 219{\cdot}0$ cm²

Try the section shown

Area provided : 2—side plates of 70 cm × 1 cm = 140 cm²
4—angles 15 cm × 11·5 cm × 1·5 cm = 4 × 37·52 = 150·08 cm²
Total area = 290·08 cm²

deduct area for holes using 22 mm diameter rivets
8 holes in angles and side plates = 8 × 2·35 × 2·5 = 47 cm²
2 holes in angles for lacing = 2 × 2·35 × 1·5 = 10·5 cm²
Total area for holes = 57·5 cm²

Net area available = 290·08 − 57·5 = 232·58 cm² > 219 cm²

Using single lacing system
$l/r \not> 170$

Use 60° inclination for the lacing,

$$\tan 60^\circ = \frac{53{\cdot}5 \times 2}{l} = \frac{2 \times 53{\cdot}5}{1{\cdot}7321} = 62 \text{ cm}$$

Using 22 mm θ for connecting width of lacing $= 2 \times 1{\cdot}5d = 3d$
$= 7$ cm

$$r_{min} = \frac{t}{\sqrt{12}}$$

$$\frac{l}{r_{min}} = 170$$

$$\frac{62}{r_{min}} = 170$$

$$r_{min} \text{ required} = \frac{62}{170} = {\cdot}365$$

Thickness required$=0{\cdot}365\times\sqrt{12}=1{\cdot}26$ cm; Use $7\times1{\cdot}6$ cm lacing bars inclined at 60°.

3·21. Diagonal Members

Max. force in the diagonal member=157·435 t (tension)

$$\text{Net area required}=\frac{157435}{1410}=112 \text{ cm}^2$$

Width of the member shall be such that it can be accommodated in the bottom and top boom members.

Use 38 cm as width of the member

Try the section shown

1—plate 38 cm × 1 cm = 38 cm²

4—angles $15\times11{\cdot}5\times1=25{\cdot}52\times4=102{\cdot}08$ cm²

Total area of the section = 140·08 cm²

Deduct for rivet holes, using 22 mm diemeter rivets

Area$=2\times3\times2{\cdot}35=14{\cdot}10$ cm²

∴ Net area of the section$=140{\cdot}08-14{\cdot}10=125{\cdot}98$ cm²
>112 cm².

3·22. Vertical Member

Maximum force in $U_2\,L_2=89{\cdot}442$ t (compression)

Length of the member = 8 m = 800 cm

Maximum width of the member = 38 cm

Select the section shown

Approximate radius of gyration; $r_{xx}=0{\cdot}39d=0{\cdot}39\times36=14{\cdot}85$ cm

$r_{yy}=0{\cdot}21\times b=0{\cdot}21\times31{\cdot}2=6{\cdot}5$ cm

$$\frac{l}{r_{min}}=\frac{800}{6{\cdot}5}=123$$

For $l/r=123$, allowable stress=609·6 kg/cm².

Area of the section; 4—angles of $15\times11{\cdot}5\times1{\cdot}2=30{\cdot}38\times4$
$=121{\cdot}52$ cm²

1—plate $38\times1=38$ cm²

Total area $=159{\cdot}52$ cm²

Safe load on the member$=159.52\times609{\cdot}6=97000$ kg>89442 kg

3·23. Connections

Join U_3. Use 22 mm diameter rivets

$$\text{Least strength in single shear}=\frac{\pi}{4}(2{\cdot}35)^2\times1020=4425 \text{ kg/cm}^2$$

Gusset plate thickness = 1 cm (should not be less than 9 mm)

Number of rivets required for member

$$U_3\,L_3=\frac{54100}{4425}=12{\cdot}2$$

Use 8 rivets on each face

Number of rivets required for member

$$U_3\,L_4=\frac{57100}{4425}=12{\cdot}9$$

Use 8 Nos on each face

Number of rivets required in $U_3\ U_4$

$$=\frac{(308{\cdot}9-279{\cdot}8)}{4{\cdot}425}=\frac{29.1}{4{\cdot}425}=6{\cdot}60. \text{ Use 8 rivets.}$$

Min spacing of rivets=2·5 d, edge distance=1·5 d

3·24. Wind Loads

Wind velocity=130 km/hr

When wind velocity exceeds 130 km/hr no live load is to be considered on the bridge

Maximum wind pressure to be considered on the unloaded bridge=240 kg/m^2

Exposed area of the structure to the wind

Top boom=42×·726 =30·5 sq. m

Bottom boom=54×0·732 =39·6 sq. m

End portals (2 Nos)=2×0·726 =13·4 sq. m

Verticals (8 Nos)=8×·31×8 =19·8 sq. m

Diagonals (8 Nos)=8×·31×9·2=22·8 sq. m

Total area=126·1 sq. m

Add 50% for the leeward truss= 63·05 sq. m

189·15 sq. m

Add for stringers, girders etc (say) 10·85 sq. m

200·00 sq. m

Total wind load=240×200=48000 kg=48 t

Wind load per portal=24 t

3.25. Analysis End Portals

Taking moments about 'C'

$H_A(2{\cdot}5+3{\cdot}35)+H_B(2{\cdot}5+3{\cdot}35)=V_A\times 8$

$H_A=H_B=12$ t

$$V_A=\frac{12{\cdot}2\times 5{\cdot}85}{8}=17{\cdot}6 \text{ t}$$

$V_B=17{\cdot}6$ t

Consider section (I)—(I)

Taking moments about F'

Force in C'E×1·25+$H_B\times 4{\cdot}6=V_B\times 2$

$$\text{Force in C'E}=\frac{17{\cdot}6\times 2-12\times 4{\cdot}6}{1{\cdot}25}=-16 \text{ t (tensio)n}$$

Force in CF Taking moments about E,

Force in CF$\times x=H_A\times 5{\cdot}85-V_A\times 4=12\times 5{\cdot}85-17{\cdot}6\times 4=0$

Force in C'F'=0

Force in FF'

Taking moments about E

Force in FF'×1·25+$V_A+4=H_A$ (2·5+3·35)

Force in FF'=0

Force in D'E

Taking moments about C', 7×5·6=Force in D'E cos $\theta\times 2{\cdot}5$

$$\text{Force in D'E}=\frac{12\times 5{\cdot}85}{\cos\theta\times 2{\cdot}5}=\frac{12\times 5{\cdot}85\times 4{\cdot}5}{4\times 2{\cdot}5}=31{\cdot}8 \text{ t (comp)}$$

Force in CE
Taking moments about D
Force in CE$\times$2·5=12$\times$3·35+24$\times$2·5

$$\text{Force in CE}=\frac{100{\cdot}25}{2{\cdot}5}=40{\cdot}25 \text{ t}$$

Member	*Force tension*	*Compression*
C′E	16 t	—
CF	0	0
C′F′	0	0
FF′	0	0
DE	—	31·8 t
D′E	—	31·8 t
CE	—	40 25 t
C′E	—	40·25

3·26. Overhead Bracing

Maximum tension occurs in $B_1 B'_2$

Force in $B_1 B'_2$=Force in $B_1 B'_1\times$Sec $\theta=24\times\frac{10}{8}=30$ t

$$\text{Net area required}=\frac{30000}{1410}=21{\cdot}4 \text{ cm}^2$$

$l/r \not> 250$

$$r_{min} \text{ required}=\frac{0{\cdot}85\times1000}{250}=3{\cdot}4 \text{ cm}$$

Using angles of 20$\times$20$\times$1·2 cm
r_{min}=3·9 cm
Areas=46·61 cm^2>21·4 cm^2
Member $B_1 B'_1$
Maximum compression=24 t
$l/r \not> 140$

$$r_{min} \text{ required}=\frac{{\cdot}85\times800}{140}=4{\cdot}85 \text{ cm}$$

ISHB$-$225, r_{min}=4·84 cm
A=54·94 cm^2
Allowable stress=494 kg/cm^2
Safe load=·494$\times$54·94=27 t>24 t

3·27. Bearings

Loads:

$$\text{Dead load reaction}=\frac{28\times8}{2}=112 \text{ t}$$

Live load reaction including impact

$$=14{\cdot}75+14{\cdot}75\left(\frac{52{\cdot}8}{54}+8{\cdot}8\times\frac{48{\cdot}5}{54}+8{\cdot}8\times\frac{45{\cdot}5}{54}+8{\cdot}8\times\frac{42{\cdot}5}{54}\right.$$
$$+8{\cdot}8\times\frac{39{\cdot}5}{54}+3{\cdot}5\times\frac{19{\cdot}5}{54}+3{\cdot}5\times\frac{18{\cdot}4}{54}+14{\cdot}75\times\frac{15{\cdot}2}{54}$$
$$\left.+14{\cdot}75\times\frac{14}{54}+8{\cdot}8\times\frac{9{\cdot}7}{54}+8{\cdot}8\times\frac{6{\cdot}7}{54}+8{\cdot}8\times\frac{3{\cdot}7}{54}+8{\cdot}8\times\frac{0{\cdot}7}{54}\right)$$

$$=14{\cdot}75+14{\cdot}45+\frac{8{\cdot}8}{54}(48{\cdot}5+45{\cdot}5+42{\cdot}5+39{\cdot}5)$$

$$+\frac{3{\cdot}5}{54}(37{\cdot}9)+\frac{14{\cdot}75}{54}(29{\cdot}2)+\frac{8{\cdot}8}{54}(20{\cdot}8)=71{\cdot}74 \text{ t}$$

Total reaction$=71{\cdot}74+126=197{\cdot}74$ t

3·27·1. Longitudinal Force

In the case of a single lane or a two lane bridge, the braking effect is 20% of the first train load plus 10% of the loads of the succeeding trains or part thereof, the train loads in one lane only being considered for the purpose of estimating the breaking effect or the longitudinal force.

Load per lane of the first train
$=2\times2{\cdot}7+2\times11{\cdot}4+4\times6{\cdot}8=55{\cdot}4$ t

20% of the train load$=55{\cdot}4\times\frac{20}{100}=11{\cdot}08$ t

Load per lane of the second train
$=2\times2{\cdot}7+2\times11{\cdot}4+2\times6{\cdot}8=41{\cdot}8$ t

10% of this load$=\frac{10}{100}\times41{\cdot}8=4{\cdot}18$ t

Total longitudinal force$=11{\cdot}08+4{\cdot}18=15{\cdot}26$ t

3·27·2. Rocker Bearing

The longitudinal force is assumed to act at 1·2 m above the road level.

Approximate position of knuckle pin from road level
$=10+20+65+70+5+15=1{\cdot}95$ m say 2 m

Vertical reaction on bearing due to longitudinal force

$$=\frac{15{\cdot}26(1{\cdot}2+2)}{54}={\cdot}960 \text{ t}=960 \text{ kg}$$

Total vertical load due to all loads transferred to bearing
$=112+71{\cdot}74+0{\cdot}96=184{\cdot}7$ t

For spans greater than 15 m on rigig pairs or abutments, bearings which will permit angular deflection of the girder ends shall be provided and at one end these shall be a rolier, rocker or other effective type if expansion bearings.

Area of sole plate or bearing plate on casting

$$=\frac{98700}{1890}=105{\cdot}7 \text{ cm}^2$$

A plate of 60 cm × 40 cm × 2·5 cm is used.

3·27·3. Casting

Knuckle pin of diameter not less than 10 cm shall be used
Use 15 cm dia pin
Allowable stress in pin$=1180$ kg/cm²

Width of rib plate required$=\frac{198700}{1180\times15}=11{\cdot}3$ cm

Use 4 ribs of 3 cm each. Total width provided$=12$ cm

3·27·4. Bending Moment in the Casting

$$\text{Bending moment}=\frac{198700}{40}\times\frac{9{\cdot}5^2}{2}=224{,}000 \text{ cm kg}$$

C.G of the arrangement

$$=\frac{60\times30\times1{\cdot}5+4\times3\times15\times10{\cdot}5}{60\times3+4\times3\times15}=\frac{270+1900}{180+180}=\frac{2170}{360}$$

$=6{\cdot}05$ cm

M.I of the section

$$=\frac{60\times3^3}{12}+60\times3\times(6{\cdot}05-1{\cdot}5)^2$$

$$+4\left[\frac{3\times15^3}{12}+15\times3\,(7{\cdot}5-3{\cdot}05)^2\right]$$

$=135+3730+4(845+890)=10805 \text{ cm}^4$

Max bending stress in the casting

$$=\frac{224000}{10805}\times11{\cdot}95=246 \text{ kg/cm}^2<1180 \text{ kg/cm}^2$$

3·27·5. Stress in pin

$$\text{Bending moment in pin}=\frac{198700\times3}{4}=149000 \text{ cm kg}$$

$$\text{Bending stress}=\frac{M}{Z}=\frac{149000}{\frac{\pi}{32}\times15^3}=450 \text{ kg/cm}^2<2120 \text{ kg/cm}$$

$$\text{Shear stress}=\frac{198700}{4\times\frac{\pi}{4}\times15^2}=268 \text{ kg/cm}^2<1020 \text{ kg/cm}^2$$

3·27·6. Rocker Design

Assume the radius of rocker$=40$ cm

Allowable stress$=0{\cdot}8\,D_r$ kg/mm length

Strength of roller$=0{\cdot}8\times2\times400=640$ kg/mm length

$=6400$ kg/cm

$$\text{Length of rocker required}=\frac{198700}{6000}=31{\cdot}1 \text{ cm}$$

Length provided$=75$ cm

Width say-30 cm

3·27·7. Steel Block

Assume 75 cm$\times$30 cm

$$\text{Bending moment}=\frac{198700}{75\times30}\times\frac{15\times15}{2}=9100 \text{ cm kg/cm width}$$

Thickness of plate required

$$\frac{b\times t^2}{6}\times f_b=\text{BM}$$

$$\frac{1\times t^2}{6}\times1890=8400$$

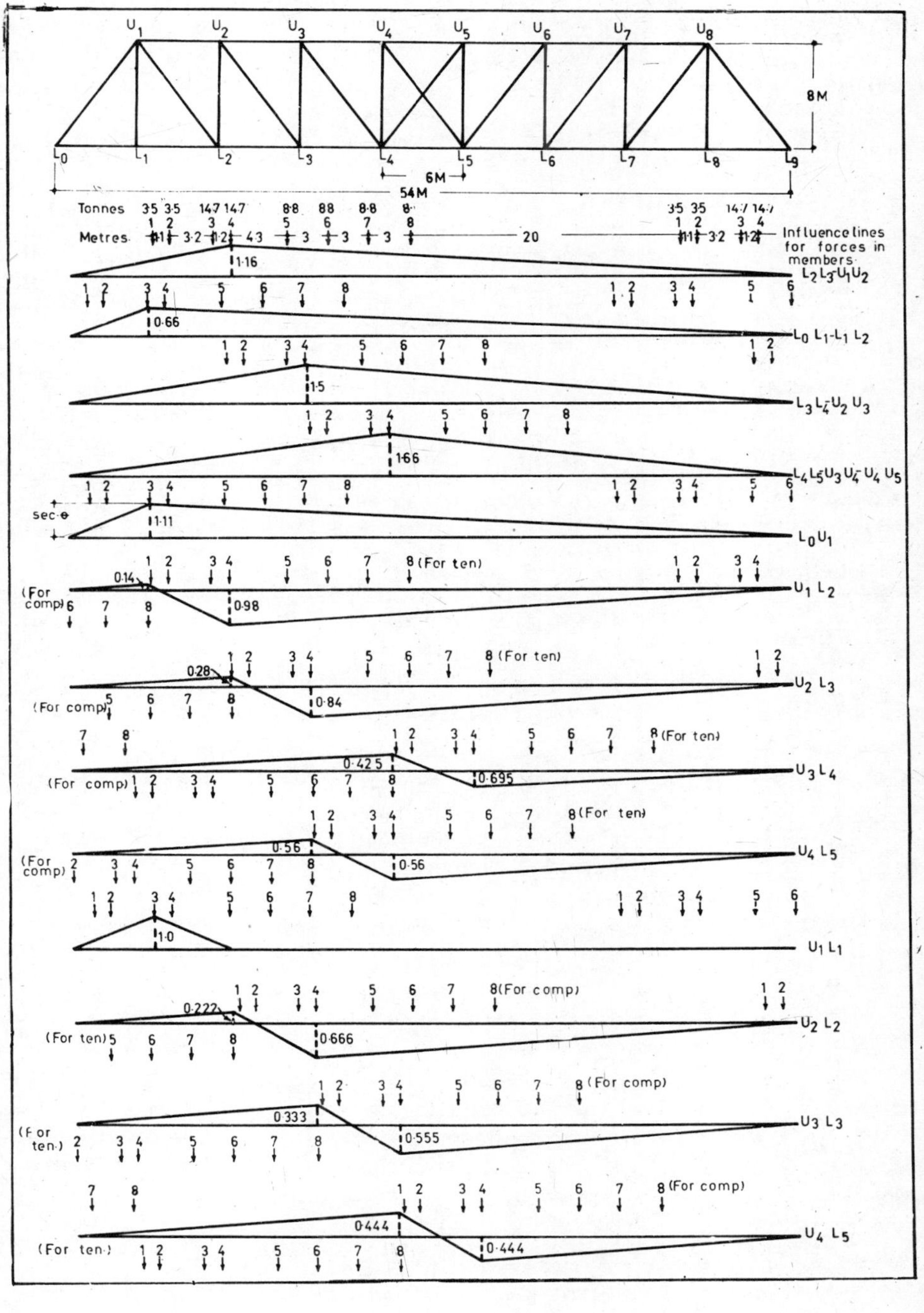
U1 U2 U3 U4 U5 U6 U7 U8
L0 L1 L2 L3 L4 L5 L6 L7 L8 L9
8M
6M
54M
Tonnes 3·5 3·5 14·7 14·7 8·8 8·8 8·8 8· 3·5 3·5 14·7 14·7
Metres 1·1 3·2 1·2 4·3 3 3 3 20 1·1 3·2 1·2
Influence lines for forces in members
1·16
L2 L3 - U1 U2
0·66
L0 L1 - L1 L2
1·5
L3 L4 - U2 U3
1·66
L4 L5 U3 U4 - U4 U5
sec θ
1·11
L0 U1
0·14
(For ten)
(For comp)
0·98
U1 L2
0·28
(For ten)
(For comp)
0·84
U2 L3
(For ten)
0·425
(For comp)
0·695
U3 L4
(For ten)
0·56
(For comp)
0·56
U4 L5
1·0
U1 L1
(For comp)
0·222
(For ten)
0·666
U2 L2
(For comp)
0·333
(For ten)
0·555
U3 L3
(For comp)
0·444
(For ten)
0·444
U4 L5

$$t=\sqrt{\frac{8400\times 6}{1890}}=5{\cdot}35 \text{ cm}$$

Use 5·5 cm thick plate

3·28. Concrete Block

Use M_{200} concrete

$$\text{Area of the concrete block}=\frac{19870}{40}=4980 \text{ cm}^2$$

Use 90 cm × 55 cm block

$$\text{Maximum bending moment}=\frac{198700}{55\times 90}\times\frac{12{\cdot}5^2}{2}$$

$$=3150 \text{ cm kg/cm length}$$

Thickness of the concrete block

$$\frac{40\times t^2\times b}{6}=3150$$

$$t=\sqrt{\frac{3150\times 6}{40\times 1}}=\sqrt{447}=21{\cdot}7 \text{ cm}$$

Use 90 cm × 50 cm × 25 cm concrete block

4. Curved Chord Steel Trussed Highway Girder Bridge—36 m Span

4·1. Data

Type = Through
Roadway= Two lane 6·8 m
Loading = IRC Class-A
Material available : Concrete= M_{150}. Steel all rolled sections.

4·2. Deck Slab

Arrangement of cross girders and stringer beams span for the main girder = 36 m

Let cross girders be placed at 6 m centre to centre and the stringer beams at 2·25 m c/c

Span for deck slab=2·25 m

l' = width of the slab
l = effective span of the slab

$$\frac{l^1}{l}=\frac{6}{2{\cdot}25}=2{\cdot}67>2$$

Slab spans in one direction only.

If a solid slab is supported on two opposite edges and carries concentrated loads, it should be designed to resist the maximum bending moment caused by the loading system. Such bending moment shall be assumed to be resisted by an effective width measured parallel to the supporting edges.

The effective width shall be calculated from the expression

$$e=Kx\left(1-\frac{x}{l}\right)+W$$

where K = a constant depending on $\frac{l^1}{l}$ ratio

x = distance of the C.G. of the concentrated load from the nearer support.

W = tyre or track contact area over the road surface of the slab in the direction at right angles to the span + twice the thickness of the wearing coat or surface finish above the structural slab.

4·3. Loads

Dead loads :

weight of slab (20 cm) $\frac{20}{100}\times1\times1\times2400=480$ kg/m² wearing coat (10 cm) $= \frac{10}{100} \times 1 \times 1 \times 2200 = 220$ kg/m^2

Total dead load $= 700$ kg/m^2

Live loads :

Class A loading system consists of

↓ 1·1 m ↓ 3·2 m ↓ 1·2 m ↓ 4·3 m ↓ 3 m ↓ 3 m ↓ 3 m ↓
2·7 t 2·7 t 11·4 t 11·4t 6·8 t 6·8 t 6·8 t 6·8t

It is assumed that maximum bending moment is produced in the deck slab when the heaviest wheel load is at the centre of span.

Therefore the distance from the nearer support to the C.G. of the concentrated load $= x = \frac{2{\cdot}25}{2} = 1{\cdot}125$ m. Effective width of the of the slab in the directions of travel of thewheel load considerd.

$$e = Kx\left(1 - \frac{x}{l}\right) + w$$

For continous $\frac{l'}{l} > 2$ from Table

$K = 2{\cdot}6$

$$e = 2{\cdot}6 \times 1{\cdot}125\left(1 - \frac{1{\cdot}125}{2{\cdot}25}\right) + 0{\cdot}25 + 2 \times 0{\cdot}10 = 1{\cdot}91 \text{ m}$$

Effective width of two wheel loads overlap, therefore the effective width $= 1{\cdot}2 + \frac{1{\cdot}91}{2} + \frac{1{\cdot}91}{2} = 3{\cdot}11$ m $<$ 6m, the width of slab, this should not however exceed the width.

Wheel load per metre width of slab $= \frac{2 \times 5{\cdot}7}{3{\cdot}11} = 3{\cdot}65$ t

4·3·1. Impact on Live Load

Impact factor for deck system upto 3 m for steel bridges $= 54{\cdot}5\%$

Live load per metre width including impact $= 3{\cdot}65 \times 1{\cdot}545 = 5{\cdot}65$ t

4·3·2. Bending Moments

Dead load bending moments

$$\text{Maximum} -ve \text{ BM} = -\frac{wl^2}{10} = -\frac{700 \times 2{\cdot}25^2}{10} \times 100$$

$$= -35500 \text{ cmkg}$$

$$\text{Maximum} +ve \text{ BM} = \frac{wl^2}{12} = \frac{700 \times 2{\cdot}25^2}{12} \times 100 = 29500 \text{ cm kg.}$$

Live load bending moment

Load dispersion from the wheel in the direction of span $= 0{\cdot}50 + 2(0{\cdot}10 + 0{\cdot}18) = 1{\cdot}06$ m

Maximum bending moment at the centre of the span

$$= \frac{5{\cdot}65}{2} \times \frac{2{\cdot}25}{2} - \frac{5{\cdot}65}{2} \times \frac{1{\cdot}06}{4} = \frac{5{\cdot}65}{2}(1{\cdot}125 - 0{\cdot}265)$$

$$= \frac{5{\cdot}65}{2} \times 0{\cdot}86 = 2{\cdot}42 \text{ mt} = 242000 \text{ cm kg}$$

Due to continuity live load bending moments can be reduced

Maximum +ve bending moment

$=0.8\times\frac{wl^2}{10}=0.8\times 242000=197.000$ cm kg

Maximum −ve bending moment

$=-\frac{wl^2}{9}=\frac{8}{9}\times 242000=-215,000$ cm kg

4·3·3. Total Bending Moment

Total +ve bending moment due to DL+LL including impact

$=29,500+197,000=226,500$ cm kg

Total −ve BM due to DL+including impact

$=35500+215000=250500$ cm kg

Maximum BM $=250,500$ cm kg

4·3·4. Shear Force

Due to DL $=0.6\times wd$

$=0.6\times 700\times 2.25=945$ kg

Maximum shear force occurs when the wheel loads are close to supports as arranged.

4·4. Effective Width For Wheel Loads (1)

Effective width for wheel load (*i*)

$e_1=kx_1\left(1-\frac{x_1}{l}\right)+W$

$x_1=0.53$

$e_1=2.6\times 0.53\left(1-\frac{0.53}{2.25}\right)+0.45=1.51$ m

Since the effective width overlap for the loads total effective width $=1.51+1.20=2.71$ m

4·4·1. Effective Width For Wheel Loads (2)

Effective width

$e_2x_2=0.08$

$e_2=2.6\times 0.08\left(1-\frac{0.08}{2.25}\right)+0.45=0.65$ m

The effective widths do not overlap

The effective width $=0.65$ m

Load per metre width due to wheel loads (1)

$=\frac{5.7\times 2}{2.71}=4.2$ t

Load per metre width due to wheel load (2)

$=\frac{5.7}{0.65}=8.8$ t

Neglecting S.F. due to LL

Continuity

$$=RA=4{\cdot}2\left(\frac{2{\cdot}25-0{\cdot}53}{2{\cdot}25}\right)+\left(\frac{8{\cdot}8\times0{\cdot}45}{1{\cdot}06}\right)\times\frac{0{\cdot}45}{2\times2{\cdot}25}$$

$=3{\cdot}2+0{\cdot}375=3.575$ t

Impact on live load$=0{\cdot}545$

S.F due to LL including impact$=1{\cdot}545\times3{\cdot}575=5{\cdot}525$ t

Maximum SF due to DL and LL including impact
$=945+5525=6470$ kg

4·5. Design of Deck Slab

Maximum BM due to all loads= 250,500 cm kg

Maximum SF due to all loads=6470 kg

Using M_{150} concrete, $8{\cdot}76\ d_e^2=250{,}500$

$$d_e^2=\frac{250500}{8{\cdot}7\times100}$$

$$d_e=\sqrt{\frac{250500}{8{\cdot}7\times100}}=17\text{ cm}$$

Cover ≮ 2·5 cm

Used $d=20$ cm $\qquad d_e=17$ cm

4·5·1. Main Reinforcement

$$A_t=\frac{250500}{1400\times0{\cdot}87\times17}=10{\cdot}55\text{ cm}^2$$

Use 12 mm ϕ at 10 cm c/c (11·31 cm²)

2×effective depth=34 cm or 30 cm

4·5·2. Secondary Reinforcement

Secondary steel is provided to resist a bending moment equal to 0·3×live load moment+0·2 times the dead load moment

$=0{\cdot}3\times215000+0{\cdot}2\times35500=64500+7100=71600$ cm kg

Using 10 mm ϕ bars

Effective depth available$=20-(2{\cdot}5+1{\cdot}2+0{\cdot}5)=15{\cdot}8$ cm

$$A_t=\frac{71600}{1400\times0{\cdot}87\times15{\cdot}8}=3{\cdot}72\text{ cm}^2$$

Use 10 mm ϕ at 20 cm c/c--(3·93 cm²)

$<3\times$effective depth$=3\times17=51$ or 45 cm

4·6. Shear Stress

$$q_s=\frac{S}{b\times a}=\frac{6470}{100\times0{\cdot}87\times17}=4{\cdot}38\text{ kg/cm}^2<5\text{ kg/cm}^2$$

4·6·1. Effective Width For Wheel Loads (3)

Load spread for load (3)

$x=0{\cdot}75$ m

$$=2{\cdot}6\times0{\cdot}75\left(1-\frac{0{\cdot}75}{2{\cdot}25}\right)+0{\cdot}45=1{\cdot}29+0{\cdot}45=1{\cdot}74\text{ m}$$

Effective widths over lap

Effective width$=1{\cdot}2\times1{\cdot}74=2{\cdot}94$ m

Load per metre width $= \frac{5{\cdot}7 \times 2}{2{\cdot}94} = 3{\cdot}88$ t

4·6·2. Shear in the Second Span

For the position of wheel loads, neglecting continuity at the support

Maximum shear force

$$= \frac{8{\cdot}8 \times 0{\cdot}61}{1{\cdot}06} \times \frac{2{\cdot}17}{2{\cdot}25} + 3{\cdot}88 \times \frac{0{\cdot}75}{2{\cdot}25} = 4{\cdot}86 + 1{\cdot}29 = 6{\cdot}15 \text{ t}$$

Shear force including impact $= 1{\cdot}545 \times 6150 = 9500$ kg
Shear force due to DL $= 945$ kg
Total shear force $= 10445$ kg

4·6·3. Shear Stress

$$\text{Shear Stress} = \frac{10445}{100 \times 0{\cdot}87 \times 17} = 7{\cdot}1 \text{ kg/cm}^2 > 5 \text{ kg/cm}^2$$

Slab fails in shear, therefore use M_{200} concrete or increase the depth at supports by providing fillets

By providing fillets of 15 cm × 15 cm
Depth available at the edge or support $= 20 + 15 = 35$ cm
Effective depth $= 32$ cm

$$\text{Shear stress} = \frac{10{,}445}{100 \times 0.87 \times 32} = 3{\cdot}75 \text{ kg/cm}^2 < 5 \text{ kg/cm}^2$$

4·7 Stringer Beam Dead Loads

Dead load due to self weight of slab and wearing coat
$= 700 \times 2{\cdot}25 = 1575$ kg/m
Dead load due to stringer (assume) $= 100$ kg/m
($d = 1/10$ to $1/12$ of span)
Total load on the stringer $= 1675$ kg/m
Maximum BM due to dead load

$$= \frac{1675 \times 6 \times 6 \times 100}{8} = 755{,}000 \text{ cm kg}$$

$$\text{Shear force due to DL} = \frac{1675 \times 6}{2} = 5025 \text{ kg}$$

To produce the maximum effect in the beam the arrangement is as shown.

4·7·1. Live Loads

Live load on stringer (2)

$$= W\frac{(2{\cdot}25 - 0{\cdot}075)}{2{\cdot}25} + \frac{W \times 0{\cdot}905}{2{\cdot}25} + \frac{W \times 0{\cdot}375}{2{\cdot}25} = 1{\cdot}55\ W$$

4·7·2. Bending Moment Due to LL

Maximum bending moment occurs under load (1) when C.G. of the loads and the load (1) are equidistant from the centre of the span.

$W = 5{\cdot}7$ t maximum wheel load

$$\text{Maximum BM} = 5{\cdot}7 \times 1{\cdot}55 \times \frac{2{\cdot}7}{6} \times 2{\cdot}7 = 214 \text{ mt}$$
$$= 2140{,}000 \text{ kg cm}$$

4·7·3. Shear Force Due to LL

The maximum SF occurs for the load position shown
Magnitude of loads (1) & (2) = 1·55×5·7 = 8·85 t
Magnitude of load (3) = 1·55×3·4 = 5·3 t
Load spread for loads (1) and (2) = 25+2×28 = 81 cm
Load spread for loads (3) = 20+2×28 = 76 cm

$$\text{Maximum SF} = 8{\cdot}85\,\frac{(6-0{\cdot}405)}{6} + 8{\cdot}85\times\frac{4{\cdot}345}{6} + \frac{5{\cdot}3\times0{\cdot}045}{6}$$

$$= 8{\cdot}25 + 6{\cdot}36 + 0{\cdot}0396 = 14{\cdot}3496 \text{ t} = 14{\cdot}35 \text{ t}$$

4·7·4. Impact Factor on LL

Impact factor for steel bridges for class *A* loading

$$= \frac{9}{13{\cdot}5+L} = \frac{9}{13{\cdot}5+6} = 0{\cdot}46$$

Maximum bending moment due to LL including impact
= 1·46×21·4 = 31·2 mt
Maximum shear force including impact
= 1·46×14·35 = 20·8 t

4·8. Design of Stringer

Maximum BM due to DL+LL including impact
= 755000+3120,000 = 3875,000 cm kg
Maximum shear force due to DL+LL including impact
= 5025+20800 = 25825 kg
Modulus of section required

$$= \frac{3875000}{1500} = 2580 \text{ cm}^3$$

Maximum available section
ISLB 600×210
Z_{xx} = 2428·9 cm³ is not sufficient
Try ISLB−450×170 with two cover plates
I_{xx} available = 1223·8 cm³

$$\text{Area of cover plate} = \frac{2580-1223{\cdot}8}{45} = \frac{1356{\cdot}2}{45} = 30 \text{ cm}^2$$

Use 30×1 cm plate

$$\text{Shear stress} = \frac{25825}{42\times{\cdot}86} = 675 \text{ kg/cm}^2 < 945 \text{ kg/cm}^2$$

Dead load check
Weight of ISLB 450 = 65·3 kg/m
Weight of plate, 30×1 cm = 25·0 kg/m < 100 kg assumed.
Total = 90·3 kg/m

4·9. Cross Girder Dead Loads

Span of cross girder = 80 m
Dead load due to slab and wearing coat = 700×6 = 4200 kg/m
This is assumed to be uniformly distributed
Dead load due to stringer = 90·3×6 = 541·8 kg
Add 10% for connections = 54·1 kg
Total = 600 kg say

This is assumed to be concentrated
Self weight of the girder (assume)=300 kg/m
(20 L+100)
Total u.d. l on the beam=4500 kg/m

4·9·1. Bending Moment Due To DL

Maximum BM at centre

$$=RA\times 4-4500\times 4\times \frac{4}{2}-600\ (4-0.625)-600\ (4-2{\cdot}875)$$

$$RA=\tfrac{1}{2}\ (4500\times 8+4\times 600)=\tfrac{1}{2}\ (36000+2400)$$

$$=\frac{38400}{2}=19200 \text{ kg}$$

$$BM=19200\times 4-37600-2020-680=76800-40300=36500 \text{ m kg}$$
$$=3650{,}000 \text{ cm kg}$$

Maximum SF due to dead load=19200 kg

4·9·2. Live Loads

Arrangement of live loads for class—A loading to obtain maximum reaction on the cross girder. 1st trial arrangement

Neglecting continuity due to other spans Reaction R_1 due to one set of wheel loads for the arrangement shown.

$$=5{\cdot}7+1{\cdot}35\times\frac{1{\cdot}7}{6}+1{\cdot}35\times\frac{2{\cdot}8}{6}+5{\cdot}7\times\frac{4{\cdot}8}{6}+3{\cdot}4\times\frac{0{\cdot}5}{6}$$

$$=5{\cdot}7+0{\cdot}63+0{\cdot}382+4{\cdot}56+0{\cdot}284=11{\cdot}556 \text{ t}$$

2nd trial arrangement

$$R_2=5{\cdot}7+1{\cdot}35\times\frac{0{\cdot}5}{6}+1{\cdot}35\times\frac{1{\cdot}5}{6}+\frac{5{\cdot}7\times 4{\cdot}8}{6}+3{\cdot}4\times\frac{1{\cdot}7}{6}$$

$$=5{\cdot}7+0{\cdot}112+0{\cdot}36+4{\cdot}56+0{\cdot}962=11{\cdot}694 \text{ t Say } 11{\cdot}7 \text{ t}$$

4·9·3. Bending Moment Due to LL

The arrangement of loads to produce maximum BM in the cross-girder is to arrange them symmetrically from centre of the span.

Maximum BM for the arrangement

$$=R\times 4-11{\cdot}7\times(4-1{\cdot}49)-11{\cdot}7\ (4-3{\cdot}29)$$

$$R=\frac{11{\cdot}7\times 4}{2}=23{\cdot}4 \text{ t}$$

$$BM=23{\cdot}4\times 4-11{\cdot}7\times 2{\cdot}51-11{\cdot}7\times 0{\cdot}71=56 \text{ mt}=5600{,}000 \text{ cm kg}$$

4·9·4. Impact Factor On LL

Impact factor for class-A loading

$$=\frac{9}{13{\cdot}5+L}=\frac{9}{13{\cdot}5+8}=0{\cdot}42$$

BM due to LL including impact

$$=1{\cdot}42\times 5600{,}000=7950{,}000 \text{ cm kg}$$

4·9·5. Shear Force

Maximum shear force will occur when the wheel load is close to the support.

$$RA\times 6=11{\cdot}7\ \{(8-1)+(8-2{\cdot}8)+(8-4{\cdot}22)+(8-6{\cdot}02)\}$$
$$=11{\cdot}7\times 17{\cdot}96$$

$$RA = \frac{11 \cdot 7 \times 17 \cdot 96}{8} = 26 \text{ t} = 26000 \text{ kg}$$

Maximum shear force including impact $= 26000 \times 1 \cdot 42$
$= 37000$ kg

4·10. Design of Cross Girder

Span $= 8$ m

Total bending moment due to DL, LL, including impact
$= 3650{,}000 + 7950{,}000 = 11600{,}000$ cm kg

Total SF due to DL, LL including impact
$= 19200 + 37000 = 56{,}200$ kg

$$\text{Modulus of section required} = \frac{11600{,}000}{1500} = 7720 \text{ cm}^3$$

Maximum size Rolled steel section provides only Z
$= 2428 \cdot 9$ cm³

Using welded connections

Area of one cover plate required in composite beam

$$= \frac{7720 - 2428 \cdot 9}{60} = \frac{5291 \cdot 1}{60} = 89 \cdot 3 \text{ cm}^2$$

Use 45×2 cm plates

4·10·1. Shear Stress

$$q_s = \frac{S}{dt_w} = \frac{56200}{60 \times 1 \cdot 05} = 890 \text{ kg/cm}^2 < 945 \text{ kg/cm}^2$$

4·11. Trussed Girder Selection of Truss

Span $= 36 \cdot 0$ m

Depth of main girder ≮ 1/10th of effective span

∴ Height of truss $= \frac{1}{5}$ to $\frac{1}{7}$ of the span $= 7 \cdot 2$ to $5 \cdot 14$ m

Use $h = 7$ m

Spacing of the main girder $= 8$ m $> 1/20$ of the span

Curved chord type of truss with cross girders at 6 m shall be used.

4·11·1. Loads

Dead loads

Due to wearing coat, deck slab, stringers and cross girders acting at each support $= 19200$ kg

Self weight of truss $= w = 15L + 550$ kg/m $= 15 \times 36 + 550$
$= 540 + 550 = 1090$ kg/m

Self weight/panel $= 1090 \times 6 = 6540$ kg

Total load at each joint $= 19 \cdot 2 + 6 \cdot 54 = 25 \cdot 74$

Say $= 26 \cdot 0$ t

Live loads Class-A loading

1.35 t		1·35 t		5·7		5·7		3·4		3·4		3·4		3·4	
↓	1·1 m	↓	3·2 m	↓	1·2 m	↓	4·3 m	↓	3 m	↓	3 m	↓	3 m	↓	20 m
(1)		(2)		(3)		(4)		(5)		(6)		(7)		(8)	
(9)		(10)		(11)		(12)		(13)		(14)		(15)		(16)	

Maximum effect is produced due to moving loads when two trains side by side cross or pass on the brldge for the arrangement shown.

For any given axle load W the load transferred to the truss

$$\frac{W}{8}(7+5\cdot2+3\cdot78+1\cdot98)=\frac{W}{8}\times17\cdot96=2\cdot24\,W$$

4·11·2. Impact Factors

Impact factor for span=36 m

$$\frac{9}{13\cdot5+36}=\frac{9}{49\cdot5}=0\cdot182$$

Live load, due to any axle load W on the truss, including impact

$=1\cdot182\times2\cdot24\,W=2\cdot65\,W$

4·11·3. Live Loads on the Truss including Impact

3·58 t		3·58 t		15·1 t		15·1 t		9·1 t		9·1 t		9·1 t		9·1 t	
↓	1·1 m	↓	3·2 m	↓	1·2 m	↓	4·3m	↓	3 m	↓	3 m	↓	3 m	↓	20 m
(1)		(2)		(3)		(4)		(5)		(6)		(7)		(8)	
(9)		(10)		(11)		(12)		(13)		(14)		(15)		(16)	

4·12. Force in Bottom Chord Members

Centre of mement is U_1.

Force in L_0 L_1

IL ordinate under U_1 for force in $L_0L_1=\dfrac{\text{BM about } U_1}{\text{perpendicular distance}}$

$$=\frac{x_1\,(l-x_1)}{l\,h_1}=\frac{3\times33}{36\times3}=\frac{11}{12}$$

IL ordinate under point $L_1=\dfrac{11}{12}\times\dfrac{30}{33}=\dfrac{5}{6}=0\cdot833$

The force in the member is maximum when the average load on the left of L_1 and average load on the Right of L_1 are equal.

that is $\dfrac{W_1}{x_1}-\dfrac{W_2}{x_2}$ should change sign

Try load (3) at L_1, $\dfrac{W_1}{x_1}=\dfrac{3\cdot58+3\cdot58+15\cdot1}{6}=\dfrac{22\cdot76}{6}=3\cdot71$

$$\frac{W_2}{x_2}=\frac{15\cdot1+9\cdot1\times4}{30}=\frac{51\cdot5}{30}=1\cdot72$$

$$\frac{W_1}{x_1}-\frac{W_2}{x_2}=3\cdot71-1\cdot7=\quad +ve$$

Shifting load (3) just right of L_1

$$\frac{W_1}{x_1}-\frac{W_2}{x_2}=\frac{7\cdot16}{6}-\frac{66\cdot2}{30}=1\cdot19-2\cdot22=-ve$$

Therefor the force in member L_0 L_1 will be maximum when load (3) is at the section L_1

Force in $L_0 L_1$

$$= \cdot 833 \left(3{\cdot}58 \times \frac{1{\cdot}7}{6} + 3{\cdot}58 \times \frac{2{\cdot}8}{6} + 15{\cdot}1 \times \frac{6}{6} + 15{\cdot}1 \times \frac{28{\cdot}8}{30} + 9{\cdot}1 \times \frac{24{\cdot}5}{30} + 9{\cdot}1 \times \frac{21{\cdot}5}{30} + 9{\cdot}1 \times \frac{18{\cdot}5}{30} + 9{\cdot}1 \times \frac{15{\cdot}5}{30} \right)$$

$$= \cdot 833\,(1{\cdot}02 + 1{\cdot}68 + 15{\cdot}1 + 14{\cdot}45 + 7{\cdot}4 + 6{\cdot}5 + 5{\cdot}6 + 4{\cdot}7)$$

$$= \cdot 833 \times 56{\cdot}45 = 47{\cdot}2 \text{ t (tension).}$$

4·12·1. Force due to Dead Load

Force in $L_0 L_1$ due to dead loads.

$$= 26 \times \frac{\cdot 833}{30}\,(30 + 24 + 18 + 12 + 6)$$

$$= 26 \times \cdot 833 \times 3 = 65{\cdot}2 \text{ t (tensile).}$$

4·12·2. Force in $L_1 L_2$

Centre of moment U_2.

Due to Live Loads

$$\text{IL ordinate under } U_2 = \frac{\text{BM about } U_2}{\text{perpendicular distance}}$$

$$= \frac{x_2\,(l - x_2)}{l\,h_2} = \frac{9 \times 27}{36 \times 5} = 1{\cdot}35$$

IL ordinate under $L_1 = 1{\cdot}35 \times \frac{6}{9} = 0{\cdot}9$

IL ordinate under $L_2 = 1{\cdot}35 \times \frac{24}{27} = 1{\cdot}2$

Force due to live load in member $L_1 L_2$

For force to be maximum in $L_1 L_2$.

$$\frac{W_1}{x_1} - \frac{W_2}{x_2} \text{ should change sign}$$

Try load (4) at L_2

$$\text{Case } (i) \quad \frac{W_1}{x_1} = \frac{3{\cdot}58 + 3{\cdot}58 + 15{\cdot}1}{12} = \frac{22{\cdot}26}{12} = 1{\cdot}85$$

$$\frac{W_2}{x_2} = \frac{15{\cdot}1 + 4 \times 9{\cdot}1}{24} = \frac{51{\cdot}5}{24} = 2{\cdot}15$$

$$\frac{W_1}{x_1} - \frac{W_2}{x_2} = 1{\cdot}85 - 2{\cdot}15 = \quad -ve$$

If load (4) is shifted to left of L_2

$$\frac{W_1}{x_1} - \frac{W_2}{x_2} = \frac{37{\cdot}36}{12} - \frac{36{\cdot}4}{24} = +ve$$

Force in $L_1 L_2$ when load (4) at L_2

$$= 3{\cdot}58 \times \left(\cdot 9 + \frac{\cdot 5}{6} \times \cdot 3 \right) + 3{\cdot}58 \left(\cdot 9 + \frac{1{\cdot}6}{6} \times \cdot 3 \right)$$

$$+ 15{\cdot}1 \left(\cdot 9 + \frac{4{\cdot}8}{6} \times 0{\cdot}3 \right)$$

$$+ 1{\cdot}2 \times 15{\cdot}1 + 9{\cdot}1 \times 1{\cdot}2 \left(\frac{19{\cdot}7}{24} + \frac{16{\cdot}7}{24} + \frac{13{\cdot}7}{24} + \frac{10{\cdot}7}{24} \right)$$

$=3{\cdot}58\times 925+3{\cdot}58\times{\cdot}98+15{\cdot}1\times 1{\cdot}14+1{\cdot}2\times 15{\cdot}1$
$+9{\cdot}1\times 1{\cdot}2\,(822+{\cdot}7+{\cdot}572+{\cdot}46)$
$=3{\cdot}32+3{\cdot}5+17{\cdot}2+18{\cdot}1+9{\cdot}1\times 1{\cdot}2\times 2{\cdot}54$
$=3{\cdot}32+3{\cdot}5+17{\cdot}2+18{\cdot}1+27{\cdot}7=69{\cdot}82$ t (tensile)

Case (ii) When load (3) at L_1

$\frac{W_1}{x_1}-\frac{W_2}{x_2}$ also changes sign

Therefore force in $L_1 L_2$ when load (3) at L_1

$$=3{\cdot}58\times\frac{1{\cdot}7}{6}\times 0{\cdot}9+3{\cdot}58\times\frac{2{\cdot}8}{6}\times 0{\cdot}9+0{\cdot}9\times 15{\cdot}1$$

$$+15{\cdot}1\left(0{\cdot}9+\frac{1{\cdot}2}{6}\times{\cdot}3\right)+9{\cdot}1\left(0{\cdot}9+\frac{5{\cdot}5}{6}\times 0{\cdot}3\right)$$

$$+9{\cdot}1\times 1{\cdot}2\left(\frac{21{\cdot}5}{24}+\frac{18{\cdot}5}{24}+\frac{15{\cdot}5}{24}\right)$$

$=0{\cdot}91+1{\cdot}5+13{\cdot}7+15{\cdot}1\times 0{\cdot}96+9{\cdot}1\times 1{\cdot}175$
$+9{\cdot}1\times 1{\cdot}2\,({\cdot}895+{\cdot}77+{\cdot}645)$
$=0{\cdot}91+1{\cdot}5+13{\cdot}7+14{\cdot}5+1{\cdot}061+9{\cdot}1\times 1{\cdot}2\times 2{\cdot}31$
$=0{\cdot}91+1{\cdot}5+13{\cdot}7+14{\cdot}5+1{\cdot}061+25{\cdot}2=56{\cdot}871$ t (tensile)

Therefore the maximum force in $L_1 L_2$ occurs when load (4) is at L_2 and $=69{\cdot}82$ t (tensile)

4·12·3. Due to Dead Load

Force in $L_1 L_2$ due to dead loads

$$=26\left({\cdot}9+1{\cdot}2+\frac{18}{24}+\frac{12}{24}+\frac{6}{24}\right)$$

$=26\,({\cdot}9+1{\cdot}2+{\cdot}75+0{\cdot}5+{\cdot}25)=26\times 3{\cdot}6=93{\cdot}5$ t (tensile)

4·12·4. Force in $L_2 L_3$

Centre of moment U_3

Influence line or divate under U_3

$$=\frac{x_3(l-x_3)}{l\times h_3}=\frac{15\times 21}{36\times 7}=1{\cdot}25$$

IL ordinate at $L_2=1{\cdot}25\times\frac{12}{15}=1{\cdot}00$

IL ordinate at $L_3=1{\cdot}25\times\frac{18}{21}=1{\cdot}07$

Try load (4) at L_3

$$\frac{W_1}{x_1}=\frac{3{\cdot}58+3{\cdot}58+15{\cdot}1+15{\cdot}1}{18}=\frac{37{\cdot}36}{18}=2{\cdot}07$$

$$\frac{W_2}{x_2}=\frac{9{\cdot}1\times 4}{18}=\frac{36{\cdot}4}{18}=2{\cdot}02$$

$$\frac{W_1}{x_2}-\frac{W_2}{x_2}=2{\cdot}07-2{\cdot}02=+\text{ve}$$

By shifting load (4) just to the right of L_3

$$\frac{W_1}{x_1}-\frac{W_2}{x_2}=\frac{22{\cdot}66}{18}-\frac{51{\cdot}5}{18}=-\text{ve}$$

Force in $L_2 L_3$

$$=3{\cdot}58\left(1+\frac{\cdot 5}{6}\times 07\right)+3{\cdot}58\left(1+\frac{1{\cdot}6}{6}\times{\cdot}07\right)$$

$$+15{\cdot}1\left(1+\frac{4{\cdot}8}{6}\times 0{\cdot}07\right)+15{\cdot}1\times 1{\cdot}07+9{\cdot}1\times 1{\cdot}07$$

$$\left(\frac{16{\cdot}7}{21}+\frac{13{\cdot}7}{21}+\frac{10{\cdot}7}{21}+\frac{7{\cdot}7}{21}\right)$$

$$=3{\cdot}58\times 1{\cdot}006+3{\cdot}58\times 1{\cdot}018+15{\cdot}1\times 1{\cdot}056$$

$$+15{\cdot}1\times 1{\cdot}07+9.1\times 1{\cdot}07\times 2{\cdot}32$$

$$=3{\cdot}6+3{\cdot}65+15{\cdot}5+16{\cdot}2+22{\cdot}6=61{\cdot}55 \text{ t (tensile)}$$

4·12·5. Due to Dead Loads

Force in $L_2\ L_3$

$$=26\left(1\times\frac{6}{12}+1+1{\cdot}07+1{\cdot}07\times\frac{12}{18}+1{\cdot}07\times\frac{6}{18}\right)$$

$$=26\ (0{\cdot}5+1+1{\cdot}07+0{\cdot}715+{\cdot}356)$$

$$=26\times 3{\cdot}641=94{\cdot}5 \text{ t (tensile)}$$

4·13. Top Chord Members Force In $L_0\ U_1$

Influence line for member $L_0\ U_1$ is the same as for the member $L_0\ L_1$. Therefore the magnitude of the forces due to Live and dead loads are the same but the nature of these forces is compression.

Therefore the maximum force in $L_0\ U_1$ due to live loads $=47{\cdot}2$ t (comp)

Due to dead loads$=65{\cdot}2$ t (comp)

4·13·1. Force in $U_1\ U_2$ Due to Live Loads

Centre of moment L_1

$$\text{IL ordinate at } L_1=\frac{\text{BM at } L_1}{\text{perpendicular distance}}$$

$$=\frac{x_1(l-x_1)}{l\times h_1}=\frac{6\times 30}{36\times 3{\cdot}8}=1{\cdot}32$$

Try load (3) at L_1 when load (3) is left of the section

$$\frac{W_1}{x_1}-\frac{W_2}{x_2}=\frac{22{\cdot}76}{6}-\frac{51{\cdot}5}{30}=3{\cdot}71-1{\cdot}72=+\text{ve}$$

When load (3) is right of section

$$\frac{W_1}{x_1}-\frac{W_2}{x_2}=\frac{7{\cdot}16}{6}-\frac{66{\cdot}6}{30}=1{\cdot}19-2{\cdot}22=-\text{ve}$$

Therefore the force in member $U_1\ U_2$ will be maximum when load (3) is at the section L_1

Force in $U_1\ U_2$

$$=1{\cdot}32\left(3{\cdot}50\times\frac{1{\cdot}7}{6}+3{\cdot}58\times\frac{2{\cdot}8}{6}+15{\cdot}1\times\frac{6}{6}+\frac{15{\cdot}1\times 28{\cdot}8}{30}\right.$$

$$\left.+9{\cdot}1\times\frac{24{\cdot}5}{30}+9{\cdot}1\times\frac{18{\cdot}5}{30}+9{\cdot}1\times\frac{15{\cdot}5}{30}\right)$$

$$=1{\cdot}32\times 56{\cdot}45=74{\cdot}5 \text{ t (comp)}$$

4·13·2. Force Due to Dead Loads

Force in $U_1 U_2$ due to dead loads.

$$=26\times\frac{1{\cdot}32}{30}(30+24+18+12+6)$$

$$=26\times1{\cdot}32\times3=103 \text{ t (comp)}$$

4·13·3. Force In $U_2 U_3$ Due to Live Loads

Centre of moment L_2

$$\text{IL ordinate at } L_2=\frac{\text{BM at } L_2}{\text{prependicular distance}}$$

$$=\frac{x_2(l-x_2)}{l\times h_2}=\frac{12\times24}{36\times5{\cdot}7}=1{\cdot}4$$

Try load (4) at L_2. When load (4) is just right of L_2

$$\frac{W_1}{x_1}-\frac{W_2}{x_2}=\frac{22{\cdot}26}{12}-\frac{51{\cdot}5}{24}=1{\cdot}85-2{\cdot}15=-\text{ve}$$

When load (4) is just left of L_2

$$\frac{W_1}{x_1}-\frac{W_2}{x_2}=\frac{37{\cdot}36}{12}-\frac{36{\cdot}4}{24}=3{\cdot}11-1{\cdot}51=+\text{ve}$$

Therefore force will be maximum when load (4) is at L_2

Force in $U_2 U_3$

$$=1{\cdot}4\left(3{\cdot}58\times\frac{6{\cdot}5}{12}+3{\cdot}58\times\frac{7{\cdot}6}{12}+15{\cdot}1\times\frac{10{\cdot}8}{12}+15{\cdot}1\right)$$

$$+9{\cdot}1\times1{\cdot}4\left(\frac{19{\cdot}7}{24}+\frac{16{\cdot}7}{24}+\frac{13{\cdot}7}{24}+\frac{10{\cdot}7}{24}\right)$$

$$=1{\cdot}4\,(1{\cdot}95+2{\cdot}26+13{\cdot}6+15{\cdot}1)$$

$$+9{\cdot}1\times1{\cdot}4\,({\cdot}822+{\cdot}7+{\cdot}572+{\cdot}446)$$

$$=1{\cdot}4\times32{\cdot}91+9{\cdot}1\times1{\cdot}4\times2{\cdot}54=46+32{\cdot}4=78{\cdot}4 \text{ t (comp)}$$

4·13·4. Force Due to Dead Loads

Force in $U_2 U_3$ due to dead loads

$$=26\times1{\cdot}4\left(\frac{6}{12}+1+\frac{18}{24}+\frac{12}{24}+\frac{6}{24}\right)$$

$$=26\times1{\cdot}4({\cdot}5+1+0{\cdot}75+0{\cdot}5+0{\cdot}25)=26\times1{\cdot}4\times3=109 \text{ t (comp)}.$$

4·13·5. Force in $U_3 U_3'$ Due to Live Loads

Centre of moment L_3

$$\text{IL ordinate at } L_3=\frac{\text{BM at } L_3}{\text{perpendicular distance}}$$

$$=\frac{x_3\,(l-x_3)}{l\,h_3}=\frac{18\times12}{36\times7}=1{\cdot}285$$

Try load (4) at L_3 when load (4) is just left of L_3

$$\frac{W_1}{x_1}-\frac{W_2}{x_2}=\frac{37{\cdot}36}{18}-\frac{36{\cdot}4}{18}=2{\cdot}07-2{\cdot}02=+ve$$

when load (4) is just right of L_3

$$\frac{W_1}{x_1}-\frac{W_2}{x_2}=\frac{22{\cdot}26}{18}-\frac{51{\cdot}5}{18}=-ve$$

Therefore the force is maximum in $U_3\ U_3'$ when Load (4) is at L_3

$$\text{Force in } U_3\ U_3'=1{\cdot}285\left(3{\cdot}58\times\frac{1{\cdot}25}{18}+3{\cdot}58\times\frac{13{\cdot}6}{18}+15{\cdot}1\times\frac{16{\cdot}8}{18}+15{\cdot}1\right)+9{\cdot}1\times1{\cdot}285\left(\frac{13.7}{18}+\frac{10{\cdot}7}{18}+\frac{7{\cdot}7}{18}+\frac{4{\cdot}7}{18}\right)$$

$$=1{\cdot}285\,(2{\cdot}48+2{\cdot}71+14{\cdot}1+15{\cdot}1)+9{\cdot}1\times1{\cdot}285\times2{\cdot}05$$

$$=1{\cdot}285\times34{\cdot}39+9{\cdot}1\times1{\cdot}285\times2{\cdot}05=44+24=68 \text{ t (comp)}$$

$$\text{Force due to dead loads}=26\times1{\cdot}285\left(\frac{6}{18}+\frac{12}{18}+1+\frac{12}{18}+\frac{6}{18}\right)$$

$$=26\times1{\cdot}285\times3=100{\cdot}7 \text{ (comp)}$$

4·14. Forces in Diagonal Members

Centre of moment is L_0

$$\text{IL ordinate under } L_1=\frac{\text{BM about } L_0}{\text{perpendicular distance}}$$

$$=\frac{\frac{5}{6}\times6}{8{\cdot}4}=\frac{5}{8{\cdot}4}=0{\cdot}595$$

Force in $U_1\ L_1$

The force in the members is maximum when the average load to the left of L_1 and the average load to the right of L_1 are equal, that is

$$\frac{W_1}{x_1}-\frac{W_2}{x_2} \text{ should change sign}$$

Try load (3) at L_1, $\dfrac{W_1}{x_1}=\dfrac{3{\cdot}58+3{\cdot}58+15{\cdot}1}{6}=\dfrac{22{\cdot}26}{6}=3{\cdot}71$

$$\frac{W_2}{x_2}=\frac{15{\cdot}1+9{\cdot}1\times4}{30}=\frac{51{\cdot}5}{30}=1{\cdot}72$$

$$\frac{W_1}{x_1}-\frac{W_2}{x_2}=3{\cdot}71-1{\cdot}72=+ve$$

shifting load (3) just to the right of L_1

$$\frac{W_1}{x_1}-\frac{W_2}{x_2}=\frac{7{\cdot}16}{6}-\frac{66{\cdot}6}{30}=1{\cdot}19-2{\cdot}22=-ve$$

Therefore the force in Member $L_0\ L_1$ will be maximum when load (3) is at the section L_1

$$\text{Force in } U_1\ L_1=0{\cdot}595\left(3{\cdot}58\times\frac{1{\cdot}7}{6}+3{\cdot}58\times\frac{2{\cdot}8}{6}+15{\cdot}1\times\frac{6}{6}+15{\cdot}1\times\frac{28{\cdot}8}{30}+9{\cdot}1\times\frac{24{\cdot}5}{30}+9{\cdot}1\times\frac{21{\cdot}5}{30}+9{\cdot}1\times\frac{18{\cdot}5}{30}+9{\cdot}1\times\frac{15{\cdot}5}{30}\right)$$

$= \cdot 595\ (1\cdot 02+1\cdot 68+15\cdot 1+14\cdot 45+7\cdot 4+6\cdot 5+5\cdot 6+4\cdot 7)$

$= 0\cdot 595 \times 56\cdot 45 = 33\cdot 6$ t (tensile)

14·14·1. Force due to Dead Load

Force in $U_1 L_1$ due to dead loads

$$= 26 \times \frac{0\cdot 595}{30}\ (30+24+18+12+6)$$

$$= 26 \times 0\cdot 595 \times 3 = 46\cdot 5 \text{ t (tensile)}$$

4·14·2. Force in $L_1 U_2$

Centre of moment is L_0

$$\text{IL ordinate at } L_1 = \frac{\text{BM about } L_0 \text{ when the unit load is at } L_1}{\text{perpendicular distance}}$$

$$= \frac{1 \times 12 - \frac{5}{6} \times 6}{10\cdot 3} = \frac{7}{10\cdot 3} = \cdot 68$$

$$\text{IL ordinate at } L_2 = \frac{\text{BM about } L_0 \text{ when the unit load is at } L_2}{\text{perpendicular distance}}$$

$$= \frac{\frac{4}{6} \times 6}{10\cdot 3} = \frac{4}{10\cdot 3} = \cdot 388$$

For maximum compressive force in $U_2 L_1$

Try load (4) at L_2, $\dfrac{W_1}{x_1} = \dfrac{3\cdot 58 + 3\cdot 58 + 15\cdot 1 \times 2}{12} = \dfrac{37\cdot 36}{12}$

$$= 3\cdot 11$$

$$\frac{W_2}{x_2} = \frac{9\cdot 1 \times 4}{24} = \frac{36\cdot 4}{24} = 1\cdot 51$$

$$\frac{W_1}{x_1} - \frac{W_2}{x_2} = 3\cdot 11 - 1\cdot 51 = +ve$$

shifting load (4) just to the right of L_2

$$\frac{W_1}{x_1} - \frac{W_2}{x_2} = \frac{22\cdot 26}{12} - \frac{51\cdot 5}{24} = 1\cdot 86 - 2\cdot 15 = -ve$$

Compressive force in $U_2 L_1$ due to live load

$$= \cdot 388 \left(\frac{\cdot 97}{2\cdot 17} \times 15\cdot 1 + 15\cdot 1 + \frac{19\cdot 7}{24} \times 9\cdot 1 + \frac{16\cdot 7}{24} \times 9\cdot 1 \right.$$

$$\left. + \frac{13\cdot 7}{24} \times 9\cdot 1 + \frac{10\cdot 7}{24} \times 9\cdot 1 \right)$$

$$- 0\cdot 68 \left(\frac{2\cdot 23}{3\cdot 83} \times 3\cdot 58 + \frac{3\cdot 33}{3\cdot 83} \times 3\cdot 58 \right)$$

$= 0\cdot 388\ (6\cdot 75 + 15\cdot 1 + 7\cdot 5 + 6\cdot 35 + 5\cdot 2 \times 4\cdot 05 - 0\cdot 68\ (2\cdot 08 + 3\cdot 1)$

$= 0\cdot 388 \times 44\cdot 95 - 0\cdot 68 \times 5\cdot 18 = 17\cdot 5 - 3\cdot 5 = 14$ t (comp)

Tensile force in $U_2 L_1$ due to live load is maximum when load (8) is at $L_1 = 0\cdot 68\ (9\cdot 1 + \frac{3}{6} \times 9\cdot 1)$

$= 0\cdot 68\ (9\cdot 1 + 4\cdot 55) = 0\cdot 68 \times 13\cdot 65 = 9\cdot 3$ t (tensile)

Force in $U_2 L_1$ due to dead loads

$$= -26\times0{\cdot}68 + 26\times0{\cdot}388\left(1+\frac{18}{24}+\frac{12}{24}+\frac{6}{24}\right)$$

$$= -26\times0{\cdot}68+26\times0{\cdot}388\ (1+\tfrac{3}{4}+\tfrac{1}{2}+\tfrac{1}{4})$$

$$= -26\times0{\cdot}68+26\times0{\cdot}388\times2{\cdot}5=-17{\cdot}6+25{\cdot}20$$

$$=7{\cdot}6 \text{ t (comp)}$$

4·14·3. Force in $L_2 U_2$

Centre of moment L_0

$$\text{IL ordinate at } L_1=\frac{\text{BM about } L_0 \text{ when the unit load is at } L_1}{\text{perpendicular distance}}$$

$$=\frac{1\times12-\frac{5}{6}\times6}{15{\cdot}5}=\frac{7}{15{\cdot}5}=0{\cdot}45$$

$$\text{IL ordinate at } L_2=\frac{\text{BM about } L_0 \text{ when the unit load is at } L_2}{\text{perpendicular distance}}$$

$$=\frac{\frac{2}{3}\times6}{15{\cdot}5}=\frac{4}{15{\cdot}5}=0{\cdot}258$$

For maximum tensile force in $U_2 L_2$ try load (4) at L_2

$\frac{W_1}{x_1}-\frac{W_2}{x_2}$ changes sign as in the case of $L_1 U_2$

$$\text{Tensile force in } U_2 L_2=0{\cdot}258\left(15{\cdot}1\times\frac{0{\cdot}97}{2{\cdot}17}+15{\cdot}1+\frac{19{\cdot}7}{24}\times9{\cdot}1+9{\cdot}1\times\frac{16{\cdot}7}{24}+9{\cdot}1\times\frac{13{\cdot}7}{24}+\frac{10{\cdot}7}{24}\times9{\cdot}1\right)$$

$$-{\cdot}45\left(\frac{2{\cdot}23}{3{\cdot}83}\times3{\cdot}58+\frac{3{\cdot}33}{3{\cdot}83}\times3{\cdot}58\right)$$

$=0{\cdot}258\times44{\cdot}95-0{\cdot}45\times5{\cdot}18=11{\cdot}60-2{\cdot}32=9{\cdot}28$ t (tensile)

Compressive force in $U_2 L_2$ due to live load is maximum when load is (8) is at L_1 and no other vehicle is present on the span.

$=0{\cdot}45\,(9{\cdot}1+\frac{3}{6}\times9{\cdot}1)=0{\cdot}45\times13{\cdot}65=6{\cdot}15$ t (comp)

Force in $U_2 L_2$ due to dead loads $=26\times0{\cdot}45-26\times0{\cdot}258\times$

$$\left(1+\frac{18}{24}+\frac{12}{24}+\frac{6}{24}\right)$$

$=11{\cdot}7-26\times0{\cdot}258\times2{\cdot}5=11{\cdot}7-16{\cdot}8=-5{\cdot}1$ t (tensile)

4·14·4. Force in L_2U_3

Centre of moment L_0

$$\text{IL ordinate at } L_2=\frac{\text{BM about } L_0 \text{ when the unit load is at } L_2}{\text{perpendicular distance}}$$

$$=\frac{1\times18-\frac{2}{3}\times6}{16{\cdot}7}=\frac{14}{16{\cdot}7}=0{\cdot}84$$

$$\text{IL ordinate at } L_3=\frac{\text{BM about } L_0 \text{ when the unit load is at } L_3}{\text{perpenditure distance}}$$

$$=\frac{\frac{1}{2}\times6}{16{\cdot}7}=0{\cdot}18$$

Try load (4) at L_3

For maximum compression in $U_3 L_2$, $\frac{W_1}{x_1} - \frac{W_2}{x_2}$ should change sign.

$$\frac{W_1}{x_1} = \frac{3{\cdot}58+3{\cdot}58+15{\cdot}1}{18} = \frac{22{\cdot}26}{18} = 1{\cdot}26$$

$$\frac{W_2}{x_2} = \frac{15{\cdot}1+9{\cdot}1\times4}{18} = \frac{51{\cdot}5}{18} = 2{\cdot}85$$

$$\frac{W_1}{x_1} - \frac{W_2}{x_2} = 1{\cdot}26 - 2{\cdot}85 = -ve$$

shifting load (4) to the left of L_3

$$\frac{W_1}{x_1} = \frac{37{\cdot}36}{18} = 2{\cdot}07$$

$$\frac{W_2}{x_2} = \frac{36{\cdot}4}{18} = 2{\cdot}02$$

$$\frac{W_1}{x_1} - \frac{W_2}{x_2} = 2{\cdot}07 - 2{\cdot}02 = +ve$$

Tensile is maximum in $L_2 U_3$ when load (4) is at L_3

$$= 0{\cdot}18\left(15{\cdot}1+9{\cdot}1\times\frac{13{\cdot}7}{18}+9{\cdot}1\times\frac{10{\cdot}7}{18}+9.1\times\frac{7{\cdot}7}{18}\times9{\cdot}1\times\frac{4{\cdot}7}{18}\right)$$

$$= 0{\cdot}18\ (15{\cdot}1+6{\cdot}95+5{\cdot}42+3{\cdot}9+2{\cdot}38)$$

$$= 0{\cdot}18\times33{\cdot}75 = 6{\cdot}1 \text{ t (tensile)}$$

Compressive force in $L_2 U_3$ due to live load is maximum when load (8) is at L_2 and no other vehicle is on the span.

Force in $L_2 U_3 = 0{\cdot}84\ (9{\cdot}1+9{\cdot}1\times\frac{9}{12}+9{\cdot}1\times\frac{6}{12}+9{\cdot}1\times\frac{3}{12})$

$$= 0{\cdot}84\times9{\cdot}1\ (1+\tfrac{3}{4}+\tfrac{1}{2}+\tfrac{1}{4}) = 0{\cdot}84\times9{\cdot}1\times2{\cdot}5 = 19{\cdot}1 \text{ t}$$

(comp)

Force in $L_2 U_3$ due to dead load $= 26\times0{\cdot}84\ (1+\frac{6}{12})$

$$-26\times0{\cdot}18\left(1+\frac{12}{18}+\frac{6}{18}\right)$$

$$= 26\times0{\cdot}84\times1{\cdot}5 - 26\times0{\cdot}18\ (1+\tfrac{2}{3}+\tfrac{1}{3})$$

$$= 26\times0{\cdot}84\times1{\cdot}5 - 26\times0{\cdot}18\times2 = 32{\cdot}8 - 9{\cdot}4 = 23{\cdot}4 \text{ t}$$

(comp)

4.14.5. Force in $L_3 U_3$

Centre of moment L_0

$$\text{IL ordinate at } L_2 = \frac{1\times18-\frac{2}{3}\times6}{22} = \frac{14}{22} = 0{\cdot}635$$

$$\text{IL ordinate at } L_3 = \frac{\frac{1}{2}\times6}{22} = \frac{3}{22} = 0{\cdot}136$$

Load (4) when placed at L_3 changes the sign as seen in the case of member $L_2 U_3$.

Compressive force is maximum when load (4) is at L_3

$$=0{\cdot}136\left(15{\cdot}1+9{\cdot}1\times\frac{13{\cdot}7}{18}+\frac{10{\cdot}7}{18}+9{\cdot}1\times\frac{7{\cdot}7}{18}+9{\cdot}1\times\frac{4{\cdot}7}{18}\right)$$

$$=0{\cdot}136\,(15{\cdot}1+6{\cdot}95+5{\cdot}42+3{\cdot}9+2{\cdot}38)=0{\cdot}136\times33{\cdot}75=4{\cdot}6 \text{ t (comp)}$$

Tensile force is maximum when load (8) is at L_2 and no other vehicle is on the span

Tensile force in $U_3\,L_3=0\ 635\,(9{\cdot}1+9{\cdot}1\times\frac{9}{12}+9{\cdot}1\times\frac{6}{12}+9{\cdot}1+\frac{3}{12}=0{\cdot}635\times9{\cdot}1\times2{\cdot}5=14{\cdot}5$ t (tensile)

Force in $L_3\,U_3$ due to dead load $=26\times0{\cdot}635\,(1+\frac{6}{12})-26\times0{\cdot}136\,(1+\frac{12}{18}+\frac{6}{18})$

$$=26\times0\ 635\times1{\cdot}5-26\times0{\cdot}136\times2=24{\cdot}7-7{\cdot}1=17{\cdot}6 \text{ t (tensile)}$$

Member	*Due to live load in tonnes*		*Due to Dead load in tonnes*		*Combined loads in tonnes*		*Design load in tonnes*
	Comp	*Tension*	*Comp*	*Tension*	*Maximum —Ten +comp*	*Maximum —Ten +comp*	
Bottom Chord Members							
$L_0\,L_1$	—	47·20	—	65·20	−112·40	−65·2	−112·40
$L_1\,L_2$	—	69·82	—	93·50	−163·32	−93·5	−163·32
$L_2\,L_3$	—	61·55	—	94·50	−156·05	−94·5	−156·05
Chord Members							
$L_0\,U_1$	47·23	—	65·20	—	+112·4	+65·2	+112 4
$U_1\,U_2$	74·50	—	103·00	—	+177·50	+103 0	+177·5
$U_2\,U_3$	78·4	—	109·00	—	+187·40	+109·0	+187·4
$U_3\,U_3'$	68·0	—	100·0	—	+168·00	+100·0	+168.00
Diagonal Members							
$U_1\,L_1$	—	33·6	—	46·5	−80·1	−46·5	−80·1
$L_1\,U_2$	14	9·3	7·6	—	+21·6	−1·7	+21·6 −1·7
$U_2\,L_2$	6·15	9·28	—	−5·1	−14·38	+1·05	−14·38 + 1 05
$L_2\,U_3$	19·1	6·10	23·4	—	+42·5	+17·3	+42 5
$L_3\,U_3$	4·6	14·5	—	17·6	−32·1	−13·0	+17·3 −32.1

4·15. Design of Truss Members Bottom Boom Member

Force in member $L_1\,L_2=163{\cdot}32$ t (tension)

$$\text{Net area required}=\frac{163320}{1410}=117{\cdot}5 \text{ cm}^2$$

Try a section consisting of
2 plates $35\times1{\cdot}5$ cm$=105$ cm^2
4 angles $7{\cdot}5\times5\times1$ cm$=46$ cm^2
Total area $=151$ cm^2

Deduct area for holes, using 22 mm diameter rivets —4 holes in angles and side plates.
$4\times2{\cdot}35\times2{\cdot}5=23{\cdot}5$ cm^2
2 holes in angles for lacing
$2\times2{\cdot}35\times1=4{\cdot}7$ cm^2
Total area $=28{\cdot}20$ cm^2
Net area provided$=151-28{\cdot}2=122{\cdot}8$ cm^2
Safe load on the member$=122{\cdot}80\times1{\cdot}41=173{\cdot}0$ t$>163{\cdot}32$ t

4·15·1. Lacing

Use single lacing system

$$\frac{l}{r} \ngtr 170$$

Using 60° inclination for the lacing

$$l=\text{distance between the lacing}=\frac{2\times38}{\tan 60^\circ}=\frac{2\times38}{1{\cdot}7321}=43{\cdot}9\text{ cm}$$

Using 22 mm ϕ rivets for connecting the lacing, width of lacing $=2\times1{\cdot}5\,d=3\,d=3\times2{\cdot}35=7{\cdot}05$ cm say 7 cm

$$r_{min}=\frac{t}{\sqrt{12}}$$

$$\frac{l}{r_{min}}=170=\frac{43{\cdot}9}{r_{min}}$$

$$r_{min}=\frac{43{\cdot}9}{170}=0{\cdot}258\text{ cm}$$

Thickness required$=0{\cdot}258\times\sqrt{12}={\cdot}89$ cm
Use 70×10 mm lacing bars inclined at 60°

4·16. Top Chord Member

Force in member $U_2\,U_3=187{\cdot}4$ t (comp)
Unsupported length of member$=640$ cm

Overall width of the member is $\nless \frac{1}{15}$ of the unsupported distance between the points of intersection of the laterel bracing or the length of compression member is $\ngtr$ 45 times the width.

4·16·1. Trial Section

$$\text{Width of the section} \nless \frac{1}{15}\times640 \nless {\cdot}42{\cdot}6\text{ cm}$$

Use 45 cm as width
Try the section consisting of
1—top flange plate$=45\times1$ cm$=45$ cm^2
2—side plates 35×1 cm$=35\times1\times2=70$ cm^2
4—angles $75\times5\times1$ cm$=4\times11{\cdot}52=46{\cdot}08$ cm^2
Total $161{\cdot}08$ cm^2
$r_{xx}=0{\cdot}4\,d=0.4\times35=14{\cdot}0$ cm

$r_{yy}=0{\cdot}34d=0{\cdot}34\times45=15{\cdot}3$ cm

Effective length $=0{\cdot}85\times640=545$ cm

$$\frac{l}{r_{min}}=\frac{545}{14}=38{\cdot}8$$

Allowable stress

for $\frac{l}{r}=30$ Pac$=1360$ kg/cm^2

for $\frac{l}{r}=40$ Pac$=1300$ kg/cm^2

$\frac{l}{r}=38{\cdot}8$ Pac$=1303$ kg/cm^2

Area required $=\frac{18740}{1303}=144$ cm^2

Area provided $=161{\cdot}08$ cm^2

4·16·2. Lacing

Force in the member $U_2\ U_3=187{\cdot}4$ t (comp)

Force in lacing $=2\frac{1}{2}\%=\frac{2{\cdot}5}{100}\times187{\cdot}4=4{\cdot}7$ t $=4700$ kg

Use single lacing system

Shear on lacing $=S=\frac{4700}{2}=2350$ kg

Spacing of lacing bars $\frac{l}{r_{min}}\not>50$

$r={\cdot}38=0{\cdot}38\times35=13{\cdot}3$ cm

$x_{yy}={\cdot}28b=0{\cdot}28\times6=1{\cdot}68$ cm

$l\not>1{\cdot}68\times50=84$ cm

or $\frac{l}{r_{min}}\not>0{\cdot}7\times$ slenderness ratio of the entire section.

$\not>0{\cdot}7\times38{\cdot}8=27{\cdot}3$ cm

$l=27{\cdot}3\times1{\cdot}68=46$ cm

use $l=45$ cm

$$\frac{l}{r}=\frac{45}{1{\cdot}68}=26{\cdot}7<50$$

Inclination of the lacing, $\tan\theta=\frac{36}{22{\cdot}5}=1{\cdot}6$

$\theta=58^\circ>50^\circ<70^\circ$

4·16·3. Force in Lacing Bar

Force in lacing $=\frac{S}{\sin\theta}=\frac{2350}{{\cdot}8480}=2780$ kg

4·16·4. Width of Lacing Bar

Using 22 mm diameter rivets for connecting the lacing, maximum width of lacing.

$=2\times$ edge distance $=3d=2{\cdot}35\times3=7{\cdot}05$ cm

Use 7·5 cm wide plates.

Actual Length of lacing bar $=\frac{36}{\sin\theta}=\frac{36}{{\cdot}5480}=42{\cdot}5$ cm

$l/r \ngtr 140$

r_{min} required$=\frac{42{\cdot}5}{140}=0{\cdot}304$ cm

r_{min} for a flat of $b \times t=\frac{t}{\sqrt{12}}$

$t=0{\cdot}304\sqrt{12}=0{\cdot}304\times3{\cdot}46=1{\cdot}05$ cm

Use 7·5 cm × 1·2 cm flat as lacing

r_{min} provided$=\frac{1{\cdot}2}{\sqrt{12}}=\frac{1{\cdot}2}{3{\cdot}46}={\cdot}346$

$$\frac{l}{r_{min}}=\frac{42{\cdot}5}{0{\cdot}346}=123<140$$

Allowable stress corresponding to $\frac{l}{r}=123$

Pac$=610$ kg/cm²

Safe load on the lacing$=7{\cdot}5\times1{\cdot}2\times610=5500$ kg>2780 kg

4·16·5. Connection

Least strength of rivet in single shear

$=\frac{\pi d^2}{4}fs=\pi\times2{\cdot}35^2\times1020=4425$ kg

Number of rivets$=\frac{2780}{4425}<1$

Use one 22 mm diameter rivet

4·16·6. Tensile Stress in Lacing Bar

Tensile stress$=\frac{2780}{(7{\cdot}5-2{\cdot}35)\,1{\cdot}2}=450$ kg/cm² <1500 kg/cm²

4·17. Diagonal Member

Maximum force in the diagonal member

$U_1 L_1=80{\cdot}1$ t (tensile)

Net area required$=\frac{80100}{1410}=57$ cm²

Width of the member shall be such that it can be accomodated in the bottom and top boom members.

Use 28 cm as width of the member.

Try 1—plate—28 × 1 = 28 cm²

4—angles of ISA 70 × 50 × 10 = 11·52 × 4 = 46·08 cm²

Total area = 74·08 cm²

Deduct for rivet holes, using 22 mm dia rivets 2 × 2·35 × 3 = 14·08 cm²

Net area provided = 60 cm²

Safe load that the member can take$=60\times1{\cdot}41=84{\cdot}5$ t$>80{\cdot}1$ t

4·17·1. Member L_2 U_3

Maximum force in L_2 $L_3=42{\cdot}5$ t (comp)

Length of member = 7·6 m = 760 cm

Width of the member = 28 cm

Try a section consisting of
1 plate -28×1 cm $=28$ cm^2
4 angles ISA $125 \times 75 \times 8$ mm $=15{\cdot}38 \times 4 = 61{\cdot}52$ cm^2
Total $= 89{\cdot}52$ cm^2
Approximate
$r_{xx} = 0{\cdot}39\ d = 0{\cdot}39 \times 28 = 10{\cdot}9$ cm
$r_{yy} = 0{\cdot}21\ b = 0{\cdot}21 \times 26 = 5{\cdot}44$ cm

$$l/r_{min} = \frac{760}{5{\cdot}44} = 139$$

for $l/r = 139$, Pac $= 500$ kg/cm.^{2}
Area provided
1 plate 28×1 cm $= 28$ cm^2
4 angles $125 \times 75 \times 8$ mm $= 4 \times 15{\cdot}38 = 61{\cdot}5$ cm^2
Total area $= 89{\cdot}52$ cm^2
Safe load on the member $= 89{\cdot}52 \times 500 = 44{\cdot}75$ t $> 42{\cdot}5$ t

4·17·2. Connections

Joint U_2
Use 22 mm diameter rivets ; least strength of rivet in single shear

$$= \pi \frac{(2{\cdot}35^2)}{4} \times 1020 = 4425 \text{ kg} = 4{\cdot}425 \text{ t}$$

Number of rivets required for member $U_2\ L_2$

$$= \frac{14{\cdot}38}{1{\cdot}425} \simeq 4$$

Use 4 rivets on each face of the member.
Number of rivets required for member $U_2\ L_1$

$$= \frac{21{\cdot}6}{4{\cdot}425} \approx 5$$

Use 5 rivets per face of the member
Number of rivets required for member $U_2\ U_3$, and $U_1\ U_2$

$$= \frac{78{\cdot}4}{4{\cdot}425} \simeq 18$$

Use 10 per face in both the members.
Similarly all other joints can be designed.

4·17·3. Wind Loads

Wind velocity $= 150$ km/hr
When wind velocity exceeds 130 km/hr. no live load is to be considered on the bridge.

Maximum wind pressure to be considered on the unloaded bridge $= 240$ kg/m^2

Exposed area of the structure to the wind.
Top boom $= 29{\cdot}7 \times 0{\cdot}35 = 10{\cdot}4$ m^2
Bottom boom $= 26 \times 0{\cdot}35 = 12{\cdot}6$ m^2
All the diagonal members $= 70{\cdot}8 \times 0{\cdot}26 = 18{\cdot}4$ m^2
Total area of windward truss $= 41{\cdot}40$ sq m
Add 50% area for the Leeward truss $= 20{\cdot}70$

Total exposed area=62·10 sq m
Total wind load=240×62·1=15000 kg
This load is assumed to act at nodal points of the wind bracing provided in the central three panels.
Load at central node of the wind bracing=5 t
Loat at end node=2·5 t
Maximum force in member AB=5 t (Comp)
Effective length of the member=400 cm
Try section ISA 200×150×10 ; $A=34$ cm²
$r_{min}=3·21$ cm

$$l/r=\frac{400}{3·21}=125<140$$

Increased allowable stress=0·634×1·33=845 kg/cm²
=0·845 t/cm²
Safe load=·845×34=28·7 t > 5 t
Force in member DE

Force in the member=5 Sec $\theta=5\times\frac{10}{8}=6·25$ t

Length of member=$\sqrt{4^2+3^2}$=5 m=500 cm
Effective length =0·8×500=400
Try ISA 200×150×10 : $A=34$ cm², $r_{min}=3·21$ cm

$$\frac{l}{r_{min}}=\frac{400}{3·21}=125<200 \text{ for tension}$$

<140 for oompression
Safe load=28·7 t>6·25 t

4·18. Bearings

Loads

Dead load reaction $=\frac{6\times 26}{2}=78$ t

Live load reaction including impact

$$=15·1+15·1\ \frac{34·8}{36}+9·1\times\frac{30·5}{36}+9·1\ \frac{27·5}{36}+9·1\times\frac{24·5}{36}$$

$$+9·1\times\frac{21·5}{36}+3·58\times\frac{1·5}{36}+\frac{3·58\times 0·4}{36}$$

=15·1+14·6+25·8+1·8=57·3 t
Total reaction=78+57·3=135·3 t

4·18·1. Longitudinal Force

In the case of single lane or two lane bridge, the breaking effect is 20% of the first train load plus 10% of the loads of the succeeding trains or part there of, the train loads in one lane only being considered for the purpose of estimating the breaking effect or the longitudinal force.

Only one full train load can be accomodated in the span. Load per lane of the train=2×2·7+2×11·4+4×6·8=55·4 t

20% of train load $=\frac{55·4\times 20}{100}=11·08$ t

The longitudinal force = 11·08 t

This force is assumed to act at 1·2 m above the road level.

4·18·2. Rocker Bearing

Approximate position of Knuckle pin from road level
$= 15 + 5 + 35 + 20 + 10 = 85$ cm. Say 1 m

Vertical reaction on bearing due to longitudinal force

$$= \frac{11·08 \times (1·2 + 1)}{36} = ·68 \text{ t}$$

Total vertical load on bearing due to all loads
$= 135·3 + 0·68 = 135·98$ t say 136 t

4·18·3. Sole Plate

$$\text{Area of sole plate} = \frac{136000}{1890} = 67 \text{ cm}^2$$

However a plate of 50 cm × 40 cm × 1·5 cm is required.

Knuckle pins of diameter not less than 10 cm shall be used

Use 15 cm diameter pin

Allowable stress in pin = 1180 kg/cm²

$$\text{width of rib plate required} = \frac{136000}{1180 \times 15} = 7·7 \text{ cm}$$

Use 3 rib of 3 cm each
width provided = 9 cm

4·18·4. Bending Moment

$$\text{Bending moment} = \frac{136000}{40} \times \frac{9·5^2}{2} = 252000 \text{ cm kg}$$

C.G. of the arrangement

$$= \frac{50 \times 2·5 \times \frac{2·5}{2} + 3 \times 15 \times 3 \times 10}{50 \times 2·5 + 3 \times 15 \times 3} = \frac{1506·5}{26} = 5·8 \text{ cm}$$

$$\text{MI of the section} = \frac{1}{12} \times 50 \times 2·5^3 + 50 \times 2·5 \times 4·55^2 + \frac{1}{12} \times 3 \times 3 \times 15^3 + 3 \times 3 \times 15 \times (4·2)^2$$

$$= 7580 \text{ cm}^4$$

$$\text{Maximum bending stress} = \frac{152000}{7850} \times 11·7 = 236 \text{ kg/cm}^2$$

4·18·5. Stresses in Pin

Bending moment in pin = Load per plate × distance between the plates of top and bottom casting.

$$= \frac{136000}{3} \times 3 = 136000 \text{ cm kg}$$

$$\text{Bending stress} = \frac{M}{Z} = \frac{13600}{\frac{\pi}{32} \times 15^3} = 410 \text{ kg/cm}^2 < 2120 \text{ kg/cm}^2$$

$$\text{Shear stress} = \frac{13\text{'}000}{3\times\frac{\pi}{4}\times 15^2} = 258 \text{ kg/cm}^2 < 1020 \text{ kg/cm}^2$$

4·18·6. Rocker Design

Use a rocker of radius = 40 cm

Allowable stress = 0 8 D_r kg/mm length

Strength of roller = 0·8 × 2 × 400 = 640 kg/mm length
= 6400 kg/cm length.

$$\text{Length of rocker required} = \frac{136000}{6400} = 21\text{·}25 \text{ cm}$$

Use a : Length of 60 cm
width say 30 cm

4·18·7. Steel Block

Use 60 cm × 30 cm

bending moment in the plate :

$$BM = \frac{136000}{60\times 30}\times 15\times\frac{15}{2} = 8500 \text{ cm kg/cm width}$$

Thickness of the plate.

$$\frac{bt^2}{6}\times f_b = BM$$

$$\frac{1\times t^2}{6}\times 1890 = 8500$$

$$t = \sqrt{\frac{8500\times 6}{1890}} = 5\text{·}21$$

Use 55 mm thick plate

4·19. Concrete Block

Use M_{200} concrete

$$\text{area of the concrete block required} = \frac{136000}{40} = 3400 \text{ cm}^2$$

Use 80 mm × 50 cm

Area provided = 4000 cm²

$$\text{Maximum bending moment} = \frac{136000}{4000}\times\frac{10^2}{2}$$

$$= \frac{136000}{4000}\times 50 = 1700 \text{ cm kg/cm}$$

$$\text{Thickness of the concrete block} = \frac{40\times t^2\times b}{6} = 1700$$

$$t = \sqrt{\frac{1700\times 6}{40\times 1}} = 16 \text{ cm length}$$

Provide 80 cm × 50 cm × 20 cm concrete block.

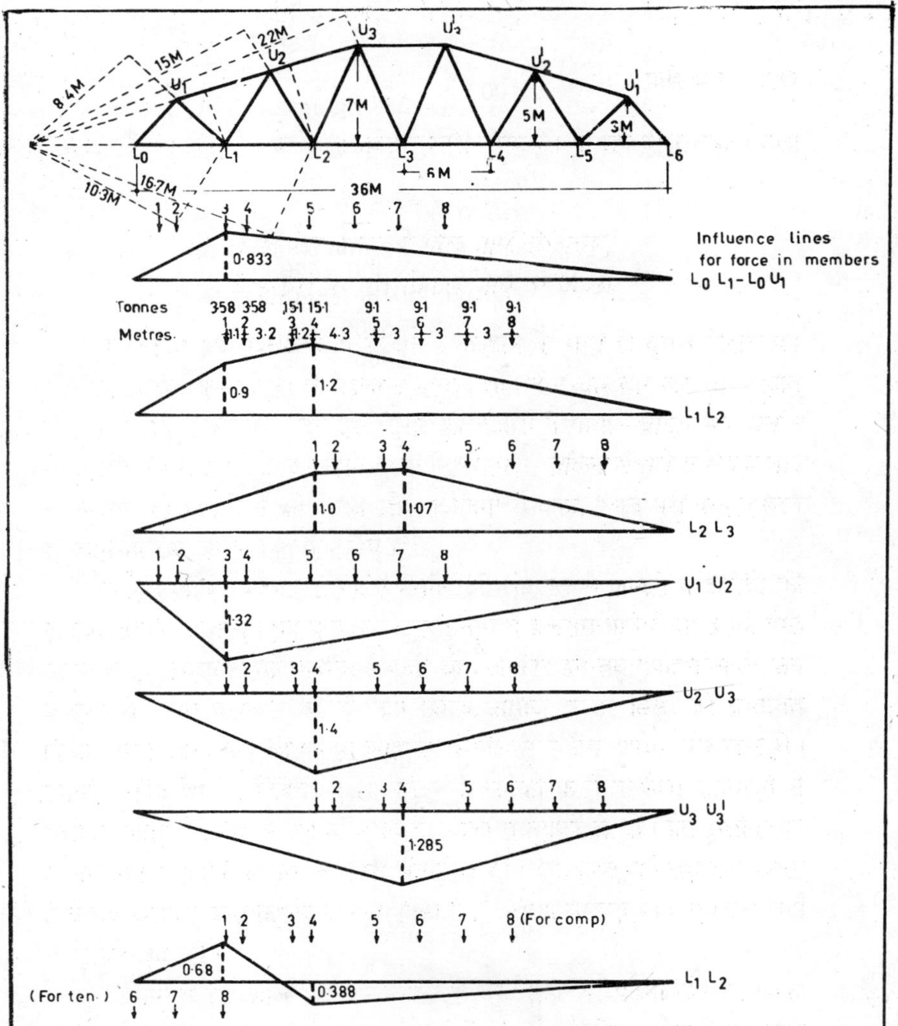

Influence lines
for force in members
$L_0 L_1 - L_0 U_1$
$L_1 L_2$
$L_2 L_3$
$U_1 U_2$
$U_2 U_3$
$U_3 U_3'$
$L_1 L_2$
$L_1 U_1$
$L_2 U_2$
$L_2 U_3$
Tonnes
Metres.
36M
6M
7M
5M
3M
8.4M
15M
22M
16.7M
10.3M
0.833
0.9
1.2
1.0
1.07
1.32
1.4
1.285
0.68
0.388
0.595
0.45
0.258
0.84
0.18
(For comp)
(For ten.)

5. Trussed Steel Highway Bridge 30 m Span

5·1. Data

Type	—	Through
Roadway	—	Two Lane
Loading	—	Class 'AA'
Safety kerbs	—	60 cm
Span	—	30 m

Material available : Concrete = M_{150} Steel = all type of rolled sections.

5·2. Deck Slab

Arrangement of cross girder and stringers

Span for cross girder = 8 m

Span for stringers = 5 m

Stringer beams are spaced at 1·7 m c/c

Span for deck slab = 1·7 m.

5·2·1. Loads

Dead loads :

weight of slab (assuming 20 cm thick

$$= \frac{20}{100} \times 1 \times 1 \times 2400 = 4{,}80 \text{ k/m}^2$$

$$\text{wearing coat (10 cm)} = \frac{10}{100} \times 1 \times 1 \times 2200 = 220 \text{ k/m}^2$$

$$\text{Total dead load} = 700 \text{ kg/m}^2.$$

Live loads : Class AA loading consists of tracked vehicle with two tracks of 3·6 m long × 0·85 m wide. Carrying a load of 35 t per track.

For maximum BM to occur, only one track will be on the span as the spacing of stringer beams is 1·7 m whereas the centre to centre of the tracks of the vehicle is 2·05 m.

5·2·2. Impact on Live Load

Impact factor for deck slabs for class 'AA' loading—Tracked vehicles 25% for spans less than 9 m.

5·2·3. Bending Moments

Dead load BM

$$\text{Maximum} - ve \text{ BM} = -\frac{wl^2}{10} = -\frac{700 \times 1{\cdot}7^2}{10} = 202{\cdot}3 \text{ mkg}$$

Maximum $+ve$ BM$=+\frac{wl^2}{12}=\frac{700\times1{\cdot}7^2}{12}=169{\cdot}0$ m kg

Dead load shear force$=700\times0{\cdot}6\times1{\cdot}7=715$ kg

Live load BM

Use pigead's method.

$U=A+2D=85+2\times10=105$ cm

$V=B+2D=360+20=380$ cm

$\frac{U}{a}=\frac{105}{170}=0{\cdot}62$

$\frac{V}{b}=\frac{380}{500}=0{\cdot}76$

$\frac{b}{a}=\frac{1{\cdot}7}{5}=0{\cdot}34$

Bending moment coefficients.

for $\rho=0{\cdot}3$ — for $\rho=0{\cdot}4$

$M_1=5{\cdot}7\times10^{-2}$ — $M_1=7{\cdot}9\times10^{-2}$

$M_2=0{\cdot}6\times10^{-2}$ — $M_2={\cdot}9\times10^{-2}$

By interpolation$={\cdot}34$

$$M_1=5{\cdot}7\times10^{-2}+\frac{(7{\cdot}9-5{\cdot}7)}{0{\cdot}1}10^{-2}\times0{\cdot}04$$

$$=(5{\cdot}7+{\cdot}88)10^{-2}=6{\cdot}58\times10^{-2}={\cdot}0658$$

$$M_2=0{\cdot}6\times10^{-2}+\frac{({\cdot}9-0{\cdot}6)}{0{\cdot}1}\times10^{-2}\times0{\cdot}04$$

$$=(0{\cdot}6+{\cdot}12)10^{-2}={\cdot}72\times10^{-2}={\cdot}0072$$

Bending moment in the transverse direction$=W(M_1+0{\cdot}15\ M_2)$

$$=\frac{70}{2}({\cdot}0658+0{\cdot}15\times{\cdot}0072)=35({\cdot}0658+{\cdot}00108)$$

$$=35\times{\cdot}06688=2{\cdot}340\text{ mt}=234000\text{ cm kg}$$

Due to continuity BM$=0{\cdot}8\times234000=187000$ cm kg

Bending moment including impact$=187000$ cm kg$\times1{\cdot}25$

$=234000$ cm kg

Bending mement in longitudinal direction$=W(M_2+0{\cdot}15\ M_1)$

$$=\frac{70}{2}({\cdot}0072+0{\cdot}15\times{\cdot}0658)=35({\cdot}0072+{\cdot}00985)$$

$$=35\ ({\cdot}01705)={\cdot}599\text{ mt}=59900\text{ cm kg.}$$

Revised bending moment due to continuity$={\cdot}8\times59900$

$=47900$ cm kg

Bending moment including impact$=1{\cdot}25\times47900=59900$ cm kg

Maximum moment in the transverse direction due to dead and live loads$=202{\cdot}3+2340=2542{\cdot}3$ m kg$=254{,}230$ cm kg

5·3. Shear Force

Load dispersion of the track in the direction of span

$=0{\cdot}85+2\ ({\cdot}20+0{\cdot}10)=0{\cdot}85+0{\cdot}6=1{\cdot}45$ m

For maximum shear to occur in the span the load is kept close to the support such that the load dispersion is within the span. That is the C.G. of the load should be at 0·725 m from support.

Effective width of slab $=Kx\left(1-\frac{x}{L}\right)+bw$

$$\frac{B}{L}=\frac{\text{width}}{\text{span}}=\frac{5}{1\cdot7}=2\cdot94$$

For $\frac{B}{L}$ ratio >2 for continuous slabs

$$K=2\cdot60$$

$$\text{Effective width}=2\cdot6\times\cdot725\left(1-\frac{\cdot725}{1\cdot7}\right)+(3\cdot6+2\times0\cdot1)$$
$$=2\cdot6\times0\cdot725\,(1-0\cdot425)+3\cdot8$$
$$=2\cdot6\times0\cdot725\times\cdot575+3\cdot8=1\cdot09+3\cdot8$$
$$=4\cdot89\text{ m}\nless5\text{ m}$$

$$\text{Load per metre}=\frac{35}{4\cdot89}=7\cdot2\text{ t}$$

$$\text{Shear force}=7\cdot2\times\frac{(1\cdot7-\cdot725)}{1\cdot7}=7\cdot2\times\frac{\cdot975}{1\cdot7}=4\cdot125\text{ t}$$

Shear force including impact at 25% $=1\cdot25\times4\cdot125=5\cdot15$ t
Shear force due to all loads $=5\cdot15+0\cdot715=5\cdot865$ t

5·4. Design of Deck Slab

Maximum SF $=5865$ kg
Maximum BM due to all loads $=254230$ cm kg
Using M_{150}, $8\cdot7\,bd_e^2=254230$

$$d_e=\sqrt{\frac{254230}{8\cdot7\times100}}=17\text{ cm}$$

Use $d=20$ cm and $d_e=17$ cm

$$A_t=\frac{254230}{1400\times\cdot87\times17}=12\cdot21\text{ cm}^2$$

5·4·1. Main Reinforcement

Use 12 mm at 9 cm c/c ($12\cdot56$ cm^2) or
16 mm at 16 cm c/c ($12\cdot57$ cm^2) <30 cm
Maximum spacing is either 30 cm or $2\times d_e=34$ cm.

5·4·2. Secondary Reinforcement

Secondary steel is provided to resist a bending moment.
$=0\cdot3\times$ live load moment $+0\cdot2$ times the dead load moment
$=0\cdot3\times234000+0\cdot2\times20230=70{,}200+4050=74250$ cm kg

Using 10 mm ϕ bars with 16 mm ϕ as main steel bars
Effective depth available $=20-(2\cdot5+1\cdot6+0\cdot5)=15\cdot4$ cm

$$A_t=\frac{74250}{1400\times\cdot87\times15\cdot4}=3\cdot97\text{ cm}^2$$

Use 10 mm ϕ at $19\cdot5$ cm c/c ($4\cdot03$ cm^2)

5·4·3. Shear Stress

$$q_s=\frac{S}{b\times a}=\frac{5865}{100\times\cdot87\times17}=3\cdot96\text{ kg/cm}^2<5\text{ kg/cm}^2$$

Span=5 m. Impact factor=25%

5·5. Stringer Beam Dead Loads

Dead load due to self weight of slab and wearing coat
$=700\times 1{\cdot}7=1190$ kg/m

Dead load due to stringer=100 kg/m (assume $\frac{1}{10}$ to $\frac{1}{12}$ of span)

Total dead load on the stringer=1290 kg/m

$$\text{Maximum BM due to DL}=\frac{1290\times 5^2}{8}=4030 \text{ m kg}$$
$$=403{,}000 \text{ cm kg}$$

$$\text{Shear force due to DL}=\frac{1290\times 5}{2}=3225 \text{ kg}$$

5·5·1. Live Loads

Stringer is subjected to maximum BM when one of the tracks is directly on it.

$$\text{Maximum BM due to LL}=\frac{35}{2}\times 2{\cdot}5-\frac{35}{2}\times\frac{3{\cdot}6}{4}$$
$$=43{\cdot}75-15{\cdot}75=28{\cdot}00 \text{ mt}=2800{,}000 \text{ cm kg}$$

BM including impact at 25%=2800,000×1·25=3500,000 cm kg

Total maximum BM due to all loads=3500,000+403,000
=3903,000 cm kg

$$Z \text{ required}=\frac{3903{,}000}{100}=2620 \text{ cm}^3$$

Use ISWB−550

Z provided=2723·9 cm^3

5·5·2. Shear Stress

Dead load SF=3·225 t

$$\text{Shear force due to LL}=\frac{35\times 3{\cdot}2}{4}=22.4 \text{ t}$$

SF due to LL including impact=22·4×1·25=28·1 t

Total SF due to all loads =28·1+3·225=31·325 t

$$\text{Shear stress}=\frac{31{\cdot}325}{1{\cdot}05\times 55}=540 \text{ kg/cm}^2<945 \text{ kg/cm}^2$$

5·6. Cross Girder (Dead load)

Span of cross girder=8m

Impact factor upto 5 m=25% and reduces to 10% upto 9 m

Dead load due to slab and wearing coat=700×5=3500 kg/m

This load is assumed to be distributed uniformly

Dead load due stringer beams=112.5×5=562·5 kg

Add for connections=38 5 kg

Total say=600 kg

This load is assumed to be concentrated

Self weight of the girder [(20L+100) kg/m)]=300 kg/m

Total uniformly distributed load=3500+300=3800 kg/m

$$R_A=\frac{600\times 5}{2}+\frac{3800\times 8}{2}=1500+15200=16700 \text{ kg}$$

Maximum BM due to dead loads$=16700\times 4-3800\times 4\times \frac{4}{2}$
$-600\times(4-0\ 6)-600(4-2{\cdot}3)$
$=66800-30400-2040-1020=33340$ m kg

Maximum SF due to DL=!6700 kg.

5·6·1. Live Load

Maximum BM occurs when the two tracks are spaced symmetrically from the centre of the cross girder

$$\text{Maximum BM}=35\times 4-35\times\frac{2{\cdot}05}{2}=140-36=104 \text{ mt}$$

Impact factor$=25\%-(25-10)\ \frac{8}{9}\%=25-13{\cdot}4=11{\cdot}6\%$.

BM including impact$=1{\cdot}116\times 104=116$ mt

Maximum SF duc to LL occurs when one of the edges of the tracks is at 1·2 m from kerb

Distance of C.G. of the load from support$=1{\cdot}2+\frac{{\cdot}85}{2}+0{\cdot}6$
$=2{\cdot}225$ m

$$SF=35\left(\frac{8-2{\cdot}225}{8}\right)+35\times\left\{\frac{8}{8}-(2{\cdot}225+2{\cdot}05)\right\}$$

$$=35\times\frac{5{\cdot}775}{8}+35\times\frac{3{\cdot}725}{8}=\frac{35}{8}(5{\cdot}775+3{\cdot}725)$$

$$=\frac{35}{8}\times 9{\cdot}5=41{\cdot}5 \text{ t}$$

Max SF including impact$=41{\cdot}5\times 1{\cdot}116=46{\cdot}2$ t

5·7. Design of Cross Girder

Span=8 m

Total maximum BM due to DL, LL including impact
=33340+11600,000=11,633,340 cm kg

Total maximum SF due to DL, LL including=16700+46200
=62900

$$Z \text{ required}=\frac{11633.340}{1500}=8450 \text{ cm}^3$$

Maximum size ISLB$-$600 rolled section provides
$Z=2428{\cdot}9$ cm^3

Using welded connections

Area of one plate required for the beam

$$=\frac{8450-2428{\cdot}9}{60}=\frac{6021{\cdot}1}{60}=103{\cdot}5 \text{ cm}^2$$

Use 45×2·5 cm plate. Area provided=113 cm^2

5·7·1. Shear Stress

$$q_s=\frac{S}{d\,t_w}=\frac{62900}{60\times 1{\cdot}05}=1000 \text{ kg/cm}^2>945 \text{ kg/cm}^2$$

Hence section is not sufficient

5·8. Plate Girder Trial Section

Try welded plate girder

$$\text{Economical depth} = 5\times\sqrt[3]{\frac{11{,}633{,}340}{1500}} = 5\times 19{\cdot}8 = 99 \text{ cm}$$

Depth required on shear consideration (assuming 10 mm thick web) $= \dfrac{62900}{1\times 945} = 66{\cdot}5$ cm

Use 80 cm × 1 cm web

$$\text{Flange area } A_f = \frac{M}{f_b\times d} - \frac{A_w}{6} = \frac{11633340}{1575\times 80} - \frac{80\times 1}{6}$$

$$= 93 - 13{\cdot}3 = 79{\cdot}7 \text{ cm}^2$$

Assuming 20 mm thick flange

$$\text{width of flange} \nless \frac{1}{50} \text{ of span}$$

$$\nless \frac{800}{50} = 16 \text{ cm}$$

$$\ngtr 32\times 2 = 64 \text{ cm}$$

Use 40 cm wide flange plate

Flange plate size$=40\times 2$ cm

$$\text{Moment of inertia } Ixx = \tfrac{1}{12}\times 1\times 80^3 + 2\times 40\times 2\,(41)^2 + \tfrac{1}{12}\times 40\times (2)^3\times 2$$

$$= 42600 + 268000 + 53 = 310653 \text{ cm}^4$$

5·8·1. Bending Stress

$$f_b = \frac{My}{I} = \frac{11633340\times 42}{310653} = 1570 \text{ kg/cm}^2 < 1575 \text{ kg/cm}^2$$

5·8·2. Connection

$$\text{Shear/cm} = v = \frac{FA\bar{y}}{I} = \frac{62900\times 40\times 2\times 41}{310653} = 660 \text{ kg.}$$

If S is the size of the weld

$$s = \frac{v}{1450} = \frac{660}{1450} = {\cdot}455 \text{ cm} = 4{\cdot}55 \text{ mm}$$

Use 5 mm size weld staggered on either side of the web plate.

Effective length of weld $\nless 10\,t = 10\times 1 = 10$ cm

Use 10 cm long welds staggered on either side.

5·8·3. Stiffners

$$\frac{d}{t} = \frac{80}{1} = 80 < 85. \text{ No intermediate stiffners are needed.}$$

However, nominal stiffners of ISA $75\times 75\times 0{\cdot}6$ under the stringer beams and at support.

5·9. Trussed Girder Selection of Truss

Span$=30$ m

Height of truss$=\frac{1}{6}$ to $\frac{1}{7}$ of span$=5$ to $4{\cdot}3$ m

Use $H=5$ m$>\frac{1}{10}$ of span

Use warren truss with 6 panels of 5 m each. Since cross girders have been used at 5 m c/c spacing of main girders=8 m>$\frac{1}{20}$ span.

5·9·1. Loads

Dead loads.

Due to wearing coat, deck slab stringers and cross girders acting at each node or at cross girder support=16700 kg=16·7 t

Self weight of truss=w=15 L+550 kg/m=15×30+550=1000 kg/m

Self weight at each nodal point=5×1000=5000 kg=5 t

Total dead load=16·7+5=21·7 t

Live loads—IRC class AA loading.

Maximum BM is produced when the class AA vehicle is closest to main girder.

Maximum load transferred when one track is at 1·625 m from the edge of kerb=41·5 t.

5·9·2. Impact Factor

For tracked vehicle 10% for all spans. Live load transferred to main girder, including impact=1·1×41·5=45·7 t.

Average uniformly distributed live load$=\dfrac{45{\cdot}7}{3{\cdot}6}=12{\cdot}7$ t/m

Live load

Forces in the members of the truss shall be determined when a live load of 12·7 t/m shorter than rolls on the bridge.

5·10. Forces in the Bottom Chord Members (Force in L_0L_1)

For force in $L_0\,L_1$

Centre of moment is U_0

$$\text{IL ordinate under } U_0=\frac{\text{BM about } U_0}{\text{perpendicular distance}}=\frac{2{\cdot}5\times27{\cdot}5}{30\times5}$$

$$=0{\cdot}46$$

$$\text{IL ordinate at } L_1=0{\cdot}46\times\frac{25}{27{\cdot}5}={\cdot}416$$

$$\frac{C_1}{p}=\frac{C}{L}$$

$$C_1=\frac{3{\cdot}6}{30}\times5=0{\cdot}6 \text{ m}$$

$$\text{ordinate at } 0{\cdot}6 \text{ m from } L_1 \text{ on the left}=\frac{y}{C_1}=\frac{{\cdot}416}{5}$$

$$y=\frac{{\cdot}416}{5}\times4{\cdot}4=0{\cdot}366$$

Force due to live load in $L_0\ L_1$

$$=\left(\frac{{\cdot}416+{\cdot}366}{2}\right)\times3{\cdot}6\times12{\cdot}7=0{\cdot}391\times3{\cdot}6\times12{\cdot}7=18 \text{ t}$$

Force due to dead load

$$=21{\cdot}7\times 0{\cdot}416\left(1+\frac{20}{25}+\frac{15}{25}+\frac{10}{25}+\frac{5}{25}\right)$$

$=21{\cdot}7\times 0{\cdot}416\times 3=27{\cdot}7$ t tension

5·10·1. Force in L_1 L_2

Centre of moment U_1

$$\text{IL ordinate at } U_1=\frac{7{\cdot}5\times 22{\cdot}5}{30\times 5}=1{\cdot}13$$

$$\text{IL ordinate at } L_1=1{\cdot}13\times\frac{5}{7{\cdot}5}=0{\cdot}75$$

$$\text{IL ordinate at } L_2=1{\cdot}13\times\frac{20}{22{\cdot}5}=1{\cdot}0$$

For maximum force to occur in L_1 L_2, section devides the load in the same ratio as it devides the span.

$$\frac{C_1}{C}=\frac{P}{L}$$

$$C_1=\frac{20}{30}\times 3{\cdot}6=2{\cdot}4 \text{ m}$$

$$\text{IL ordinate at } 2{\cdot}4 \text{ m}=\frac{1{\cdot}0}{20}\times 17{\cdot}6=0{\cdot}88$$

IL otdinate at 1·2 m left of section

$$=0{\cdot}75+\frac{3{\cdot}8}{5}\times 0{\cdot}25={\cdot}75+0{\cdot}19={\cdot}94$$

Force due to live load in L_1 L_2

$$=\frac{(0{\cdot}88+0{\cdot}94)}{2}\times 3{\cdot}6\times 12{\cdot}7={\cdot}91\times 3{\cdot}6\times 12{\cdot}7=41{\cdot}7 \text{ t}$$

Force due to dead load

$$=21{\cdot}7\left(0{\cdot}75+1+\frac{15}{20}+\frac{10}{20}+\frac{5}{20}\right)=21{\cdot}7\times 3{\cdot}25=70{\cdot}5 \text{ t}$$

Force in L_2 L_3

Centre of moment U_2

$$\text{IL ordinate at } U_2=\frac{12{\cdot}5\times 17{\cdot}5}{30\times 5}=1{\cdot}46$$

$$\text{IL ordinate at } L_2=1{\cdot}46\times\frac{10}{12{\cdot}5}=1{\cdot}17$$

$$\text{IL ordinate at } L_3=1{\cdot}46\times\frac{15}{17{\cdot}5}=1{\cdot}26$$

Considering section L_3 for max. force the section devides the load in the same ratio as it devides the span.

$$\frac{C}{3{\cdot}6}=\frac{15}{30} \text{ i.e. } C=3{\cdot}6\times\frac{15}{30}=1{\cdot}8 \text{ m}$$

IL ordinate at 1·8 m right of $L_3 = 1{\cdot}26 \times \frac{13{\cdot}2}{15} = 1{\cdot}11$

IL ordinate at 1·8 m left of L_3

$= 1{\cdot}17 + \frac{(1{\cdot}26 - 1{\cdot}17)}{5} \times 3{\cdot}2 = 1{\cdot}13$ Say

Force due to LL $= \frac{(1{\cdot}13 + 1{\cdot}11)}{2} \times 3{\cdot}6 \times 12{\cdot}7 = 51{\cdot}1$ t

Force due to DL

$= 21.7 \left\{ (1{\cdot}17 + 1{\cdot}26 + \frac{1{\cdot}7 \times 5}{10} + 1{\cdot}26 \left(\frac{10}{15} + \frac{5}{15} \right) \right\}$

$= 21{\cdot}7\ (1{\cdot}17 + 1{\cdot}26 + 0{\cdot}585 + 1{\cdot}26) = 4{\cdot}275 \times 21{\cdot}7 = 93$ t

5·11. Top Chord Members (Force in $U_0\ U_1$)

Centre of moment L_1

IL ordinate for force at $L_1 = \frac{\text{BM at } L_1}{\text{perpendicular distance}}$

$= \frac{5 \times 25}{30 \times 5} = \frac{5}{6} = {\cdot}833$

For max force in $U_0\ U_1$, consider section L_1. Section devides the load in the same ratio as it devides the span.

$C = 3{\cdot}6 \times \frac{5}{30} = 0{\cdot}6$ m

IL ordinate to the left of $L_1 = {\cdot}833 \times \frac{4{\cdot}4}{5} = {\cdot}735$

Force due to LL $= \frac{({\cdot}735 + 0{\cdot}833)}{2} \times 12{\cdot}7 \times 3{\cdot}6$

$= {\cdot}784 \times 12{\cdot}7 \times 3{\cdot}6 = 36$ t (comp)

Force due to DL $= 21{\cdot}7 \times 0{\cdot}833 \left(1 + \frac{20}{25} + \frac{15}{25} + \frac{10}{25} + \frac{5}{25} \right)$

$= 21{\cdot}7 \times 0{\cdot}833 \times 3 = 54{\cdot}1$ t

5·11·1. Force in $U_1\ U_2$

Centre of moment L_2

IL ordinate for force at $L_2 = \frac{10 \times 20}{30 \times 5} = \frac{4}{3} = 1{\cdot}33$

For maximum force in $U_1\ U_2$ consider section L_2.

Section desides the load in the same ratio as it devides the span.

$0 = 3{\cdot}6 \times \frac{10}{30} = 1{\cdot}2$ m

IL ordinate at 1·2 m left of section $= 1{\cdot}33 \times \frac{8{\cdot}8}{10} = 1{\cdot}17$

Force due to LL $= 3{\cdot}6 \times 12{\cdot}7 \times \frac{(1{\cdot}17 + 1{\cdot}33)}{2}$

$= 3{\cdot}6 \times 12{\cdot}7 \times 1{\cdot}25 = 57$ t (comp)

Force due to DL $=21\cdot7\times1\cdot13\left(\frac{5}{10}+1+\frac{15}{20}+\frac{10}{20}+\frac{5}{20}\right)$

$=21\cdot7\times1\cdot13\times3=73\cdot5$ t comp.

5·11·2. Force in U_2 U_3

Centre of moment L_3

IL ordinate for force at $L_3=\frac{15\times15}{30\times5}=1\cdot5$

For maximum force in U_2 U_3, consider section L_3

$C=3\cdot6\times\frac{15}{30}=1\cdot8$ m

IL ordinate at 1·8 m $=1\cdot5\times\frac{13\cdot2}{15}=1\cdot32$

Force due to LL $=12\cdot7\times3\cdot6\times\frac{(1\cdot5+1\cdot32)}{2}$

$=12\cdot7\times3\cdot6\times1\cdot41=64\cdot6$ t

Force due fo DL $=21\cdot7\times1\cdot5\left(\frac{5}{15}+\frac{10}{15}+1+\frac{10}{15}+\frac{5}{15}\right)$

$=21\cdot7\times1\cdot5\times3=98$ t

5·12. Forces in Diagonal Members (Force in L_0 U_0)

Force in L_0 U_0 is governed by the shear Force in the panel L_0 L_1

IL ordinate at $L_1=\frac{30-5}{30}$ Sec θ

$=\frac{25}{30}\times1\cdot13=0\cdot94$

Force due to DL $=21\cdot7\times0\cdot94\left(1+\frac{20}{25}+\frac{15}{25}+\frac{10}{25}+\frac{5}{25}\right)$

$=21\cdot7\times0\cdot94\times3=61\cdot5$ t comp

Force due to DL

Maximum compressive force will occur when head of the load touches the section

IL ordinate at 3·6 m from the section on the right

$=\left(\frac{25-3\cdot6}{25}\right)\times0\cdot94=\frac{21\cdot4}{25}\times0\cdot94=0\cdot8$

Force $=\frac{1}{2}(0\cdot94+0\cdot8)\times3\cdot6\times12\cdot7=39\cdot6$ t (comp)

5·12·1. Force in U_0 L_1

IL is same as per member L_0 U_0

Force due to DL $=61\cdot5$ tension

Force due to $L=39\cdot6$ t tension

Force in the member L_1 U_1 is governed by shear Force in panel L_1 L_2

IL ordinate at $L_1=\frac{5}{30}\times 1{\cdot}13=0{\cdot}19$

IL ordinate at $L_2=\frac{20}{30}\times 1{\cdot}13=0{\cdot}755$

Force due to DL

$$=-0{\cdot}19\times 21{\cdot}7+0{\cdot}755\times 21{\cdot}7\left(1+\frac{15}{20}+\frac{10}{20}+\frac{5}{20}\right)$$

$=-4{\cdot}13+0{\cdot}755\times 21{\cdot}7\times 2{\cdot}5$
$=-4{\cdot}13+41=36{\cdot}87$ t (comp)

Force due to LL
Maximum compressive force
$=\frac{1}{2}\times 0{\cdot}72\times 3{\cdot}6\times 12{\cdot}7=16{\cdot}8$ t
Maximum Tensile force
$=\frac{1}{2}\times 0{\cdot}19\times 1{\cdot}22\times 12{\cdot}7+\frac{1}{2}(0{\cdot}09+0{\cdot}19)\times 2{\cdot}38\times 12{\cdot}7$
$=1{\cdot}47+6{\cdot}40=7{\cdot}87$ t

5·12·2. Force in Member U_1 L_2

Force is governed by the shear force in the panel L_1 L_2
Influence line is the same as per member L_1 U_1
Force due to DL=36·87 t tension
Maximum Tensile force=16·8 t
Maximum compressive force=7·88 t

5·12·3. Member L_2 U_2

Force in the member is governed by shear force in panel L_2 L_3

IL ordinate at $L_2=1{\cdot}13\times\frac{10}{30}=0{\cdot}377$

$L_3=1{\cdot}13\times\frac{15}{30}=0{\cdot}565$

Maximum compressive Force

$$=\frac{1}{2}\times 3\times{\cdot}565\times 12{\cdot}7+\left(\frac{{\cdot}565}{2}+{\cdot}542\right)0{\cdot}6\times 12{\cdot}7$$

$=10{\cdot}8+4{\cdot}22=15{\cdot}02$ t

Maximum Tensile force

$$=\frac{1}{2}\times{\cdot}377\times 2\times 12{\cdot}7+\left(\frac{{\cdot}316+{\cdot}377}{2}\right)\times 1{\cdot}6\times 12{\cdot}7$$

$=4{\cdot}8+{\cdot}347\times 1{\cdot}6\times 12{\cdot}7=4{\cdot}8+7{\cdot}05=11{\cdot}85$ t

Force due to DL

$$=-21{\cdot}7\times{\cdot}377\left(1+\frac{5}{10}\right)+{\cdot}565\times 21{\cdot}7\left(1+\frac{10}{15}+\frac{5}{15}\right)$$

$=12{\cdot}3$ t comp

5·12·4. Force in Member U_2 L_3

IL is the same as per the member L_2 U_2.
Force due to DL=12·3 t tension
Maximum compressive force=11·85 t
Maximum Tensile force=15·02 t

Member	Due to LL in tonnes		Due to DL in tonnes		Combined loads in tonnes		Design loads in tonnes
	Comp	Tension	Comp	Tension	Max −Ten +Comp	Min −Ten +Comp	
Bottom Chord Members							
L_0L_1	—	18·0	—	27·7	−45·7	−27·7	−45·7
L_1L_2	—	41·7	—	70·5	−112·2	−70·5	−112·2
L_2L_3	—	51.1	—	93·0	−144.1	—93	−144·1
Top Chord Members							
U_0U_1	36	—	54·1	—	+90·1	+54·1	+90·1
U_1U_2	57	—	73·5	—	+130·5	+73·5	+130·5
U_2U_3	64.6	—	98	—	+162 6	+98	+162·6
Diagonal Members							
L_0U_0	39·6	—	61·5	—	+101·1	+61·5	+101·1
U_0L_1	—	39·6	—	61·5	−101·1	−61·5	−101·1
L_1U_1	16·8	7·87	38·87	—	−7·87	+38·87	−7·87 +55·67
U_1L_2	7·87	16 8	—	38·87	+55·67	−55·67	+7 87
L_2U_2	15·02	11·85	12·3	—	−55·67	+27·32	−55·67
U_2L_3	11·85	15·02	—	12·3	+7·87	−11·85 −27·32 +11·85	

5·13. Design of Members (Bottom Boom Member)

Force in member $L_2L_3 = 144{\cdot}1$ t (tension)

Net area required $= \dfrac{144100}{1440} = 102{\cdot}5$ cm²

Try a section consisting of

2—plates $35 \times 1{\cdot}2$ cm $= 84$ cm²

4—angles $7{\cdot}5 \times 5 \times 1$ cm $= 46$ cm²

Total area $= 130$ cm²

Deduct area for holes in angles and side plates

4—holes of 2·35 dia (using 22 mm ϕ rivets)

$= 4 \times 2{\cdot}35 \times 2{\cdot}2$ $= 20{\cdot}6$ cm²

2—holes in angles for lacing $= 2 \times 2{\cdot}35 \times 1 = 4{\cdot}7$ cm²

Total area $= 25{\cdot}3$ cm²

Net area provided $= 130 - 25{\cdot}3 = 104{\cdot}7$ cm²

Safe load that the trial section can carry $= 104{\cdot}7 \times 1{\cdot}41$

$= 148 \text{ t} > 144$ t.

5·14. Top Boom Members

Force in member U_2U_3=162·6 t (comp)
Unsupported length of the member=500 cm
Try a section consisting of

1—top flange plate=50×0·8=40 0 cm²
2—ISMC 350=2×53·66=107·32 (147·32 cm²)

$r_{xx}=0{\cdot}4d=0{\cdot}4\times35=14$ cm
$r_{yy}=0{\cdot}34d=3{\cdot}34\times50=17{\cdot}0$ cm
Effective length=0·85×500=·425 cm₂

$$\frac{l}{r_{min}}=\frac{425}{14}=30{\cdot}2$$

for $l/r=30 \quad P_{ac}=1360 \text{ kg/cm}^2$
$l/r=40 \quad P_{ac}=1300 \text{ kg/cm}^2$

$$\frac{l}{r}=30{\cdot}2 \; P_{ac}=1360-\frac{60\times{\cdot}2}{10}$$

$$=1360-1{\cdot}2=1358{\cdot}8 \text{ kg/cm}^2$$

$$\text{Area required}=\frac{162600}{13588}=121 \text{ cm}^2$$

Area provided=147·32 cm²

Lacing can be designed in the same way as in the previous design. Use 7·5 cm×1·2 cm flat as lacing.

5·14·1. Lacing

Use single lacing system $\frac{l}{r} \not> 170$

Using 60° inclination for the lacing

$$l=\text{distance betweeu the lacing}=\frac{2\times30}{\tan 60°}=\frac{2\times38}{1{\cdot}7321}$$
$$=43{\cdot}9 \text{ cm}$$

Using 22 mm ϕ rivets for connecting the lacing
width of lacing=2×1·5d=3d=3×2·35=7·05 cm
say 7 cm

$$r_{min}=\frac{t}{\sqrt{12}}$$

$$l/r_{min}=170=\frac{43{\cdot}9}{r_{min}}$$

$$t_{min}=\frac{43{\cdot}9}{170}=0{\cdot}258 \text{ cm}$$

Thickness required=0·258×$\sqrt{12}$=0·89 cm

Use 70×10 mm lacing bars inclined at 60°.

5·15. Diagonal Members

Force in member U_0L_1=101·1 t (tension)

$$\text{New area required}=\frac{101100}{1410}=72 \text{ cm}^2$$

Width of the member shall be such that it can be accomodated in the bottom and top booms.

Use width of the member = 28 cm

Try: 1 – plate $28 \times 1{\cdot}5 = 42{\cdot}0\ cm^2$

4 – angles ISA $7{\cdot}5 \times 7{\cdot}5 \times 1 = 4 \times 14{\cdot}02 = 56{\cdot}08\ cm^2$

Total $= 98{\cdot}08\ cm^2$

Deduct for rivet

Using 22 mm ϕ rivets

$2 \times 2{\cdot}35 \times 3{\cdot}5 = 16{\cdot}4\ cm^2$

Net area provided $= 98{\cdot}02 - 16{\cdot}40 = 81{\cdot}68\ cm^2$

Safe load that the member can take $= 81{\cdot}68 \times 1{\cdot}41 = 115$ t

Force in $L_1U_1 = 55{\cdot}67$ t comp

$-7{\cdot}87$ tension

Length of member = 560 cm

Width of member = 28 cm.

Try a section consisting of 1 plate $28 \times 1 = 28\ cm^2$

4 angles $75 \times 75 \times 8 = 4 \times 15{\cdot}38 = 61{\cdot}52\ cm^2$

$= 89{\cdot}52\ cm^2$

$0{\cdot}39 \times d = 0{\cdot}39 \times 28 = 10{\cdot}9$ cm

$0{\cdot}21 \times b = 0{\cdot}21 \times 26 = 5{\cdot}44$ cm

$$\frac{l}{r} = \frac{560}{5{\cdot}44} = 103$$

for $\frac{l}{r} = 103 \quad P_{ac} = 779\ kg/cm^2$

Safe load that the member can carry $= 0{\cdot}779 \times 89{\cdot}52 = 69{\cdot}5$ t

$> 55{\cdot}67$ t.

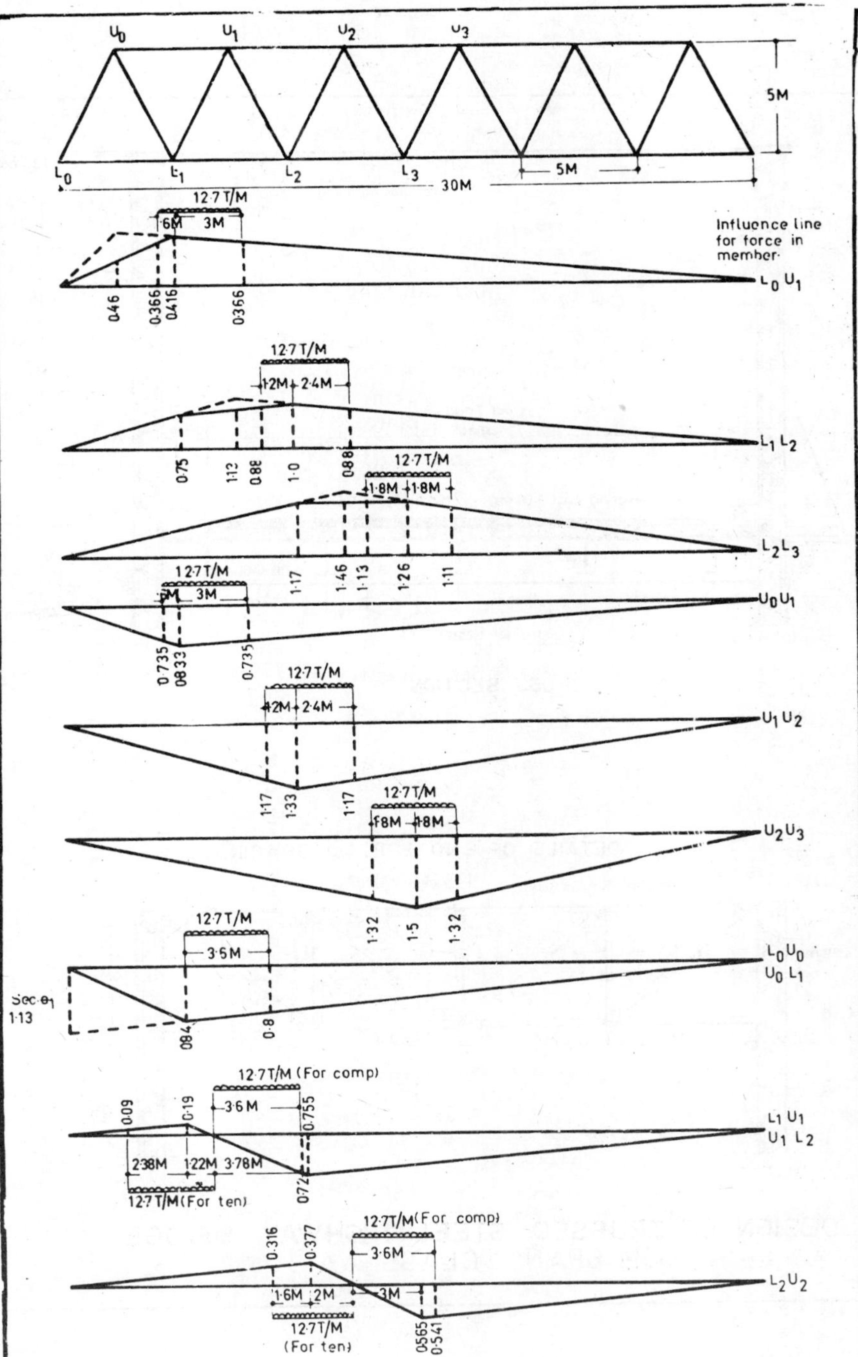

U0
U1
U2
U3
5M
L0
L1
L2
L3
5M
30M
12·7 T/M
·6M
3M
Influence line for force in member
L0 U1
0·46
0·366
0·416
0·366
12·7 T/M
1·2M
2·4M
L1 L2
0·75
1·12
0·88
1·0
0·88
12·7 T/M
1·8M
1·8M
L2 L3
1·17
1·46
1·13
1·26
1·11
12·7 T/M
·3M
3M
U0 U1
0·735
0·833
0·735
12·7 T/M
1·2M
2·4M
U1 U2
1·17
1·33
1·17
12·7 T/M
1·8M
1·8M
U2 U3
1·32
1·5
1·32
12·7 T/M
3·6M
L0 U0
U0 L1
Sec θ1
1·13
0·84
0·8
12·7 T/M (For comp)
3·6M
0·09
0·19
0·755
2·38M
1·22M
3·78M
0·72
12·7 T/M (For ten)
L1 U1
U1 L2
12·7 T/M (For comp)
0·316
0·377
3·6M
1·6M
2M
3M
0·565
0·541
12·7 T/M
(For ten)
L2 U2

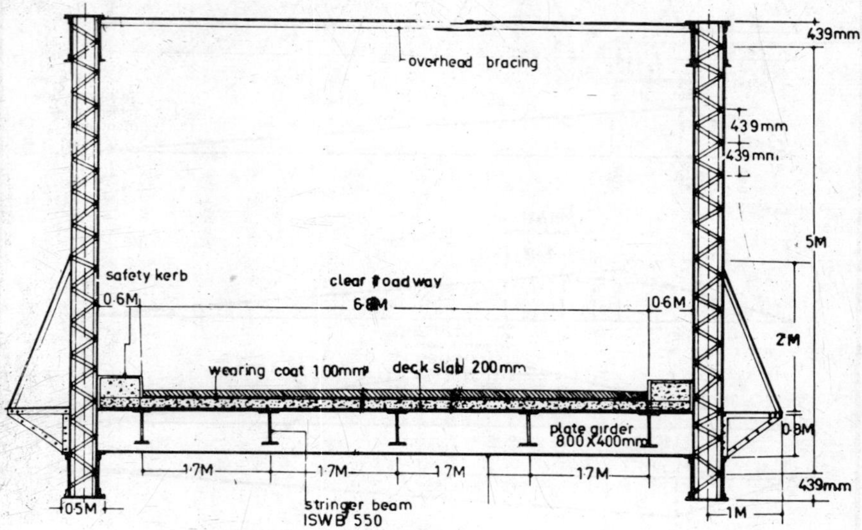

CROSS SECTION

DETAILS OF END PORTAL BRACING

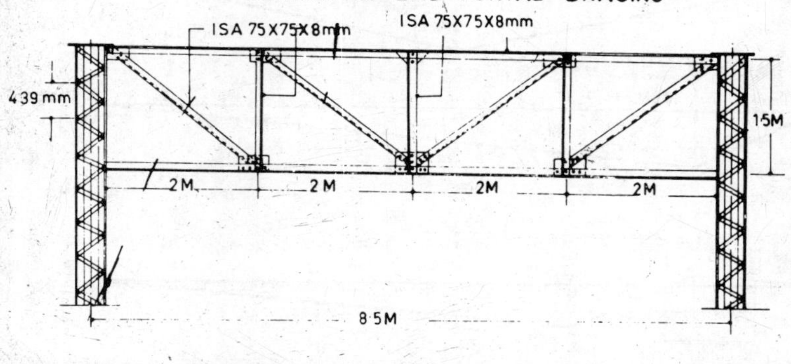

DESIGN OF TRUSSED STEEL HIGHWAY BRIDGE
30M SPAN CLASS AA

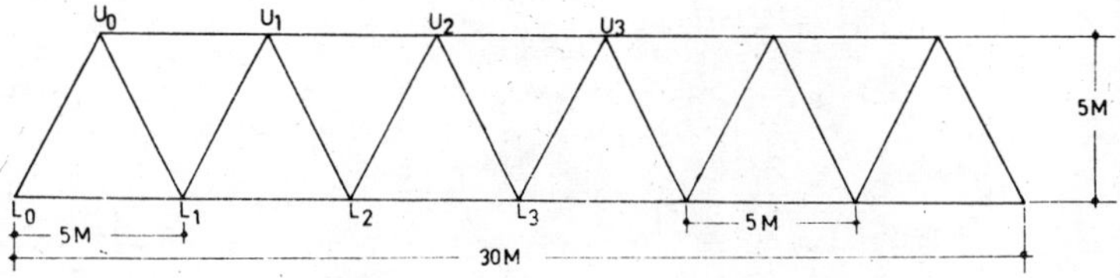

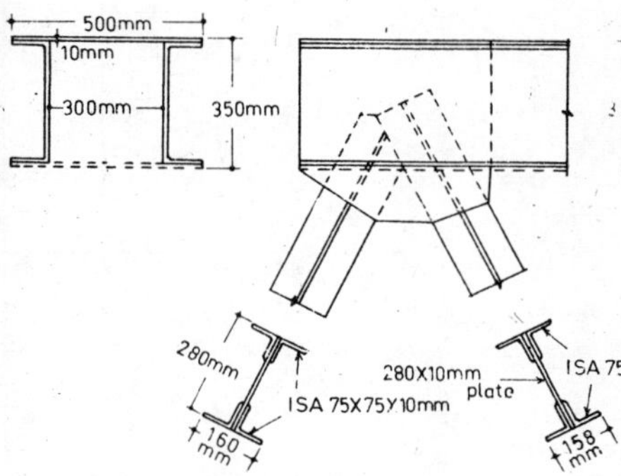

DETAIL AT U_0

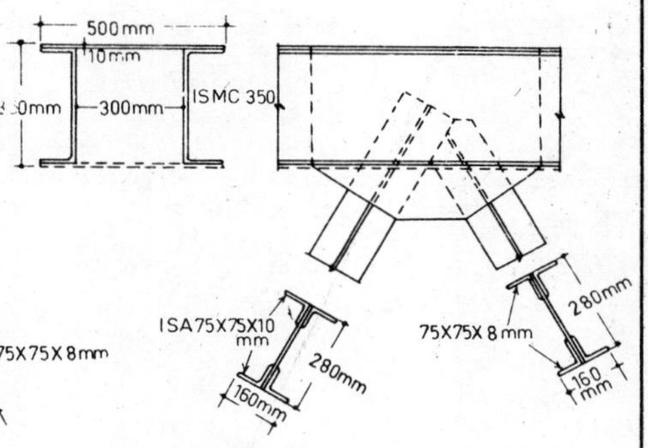

DETAIL AT U_2

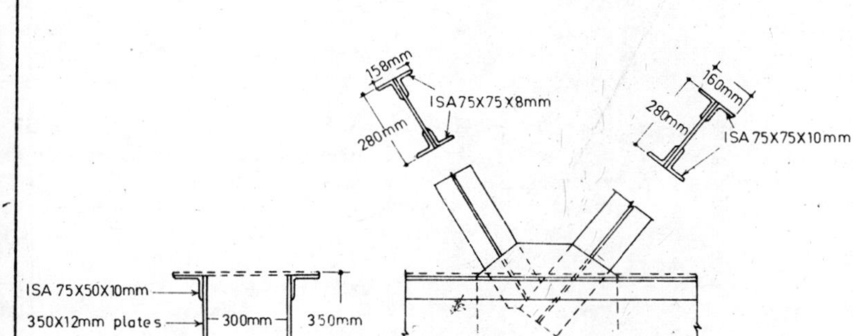

DETAIL AT L_2

PART III

DESIGN OF RAILWAY BRIDGES

1. Railway Bridges—General

1·1. Type

1. Foot bridges
2. Rolled beam bridges
3. Plate girder bridges
4. Trussed girder bridges.

1·2. Loading

These bridges are designed to carry either Main line or Branch line loading. The loading would depend upon the type and number of tracks it has to carry namely Broad gauge, metre gauge or Narrow gauge.

1·2·1. Dead loads

For a single track weight of rails, inside guide rails
and fastenings $=300$ kg/m
weight of ballast$=1600$ kg/m^3
Total weight of open floor where 254 mm$\times$254 mm$\times$2795 mm
wooden sleepers at 350 mm c/c are used$=750$ kg/m
Self weight of plate girders/m for 10 m to 30 m spans
$=w=20L+100$ kg
Self weight of trussed bridges for 30 m to 90 m spans
$=w=15\,L+550$ kg.

1·2·2. Live Loads Foot Bridges

Live load$=490$ kg/m^2
Impact is not considered for foot bridges.

1·3. Railway Bridges

Railway brides shall be designed for one of the following standards of railway loading.

(*i*) Standard main line (ML) and Branch line (BL)
Broad gauge (1676 mm)
Equivalent uniformly distributed loads on each track for calculating bending moments and shear forces are specified by the Indian Railway Board for different spans.

(*ii*) Standard Main Line (ML), Standard Branch Line (BL) and standard C. for metre gauge (1000 mm) *E, U, d, l* on each track for computing BM and SF are specified by the Indian Railway Board.

1·4. Impact Factors On live loads

(*i*) For single track

$$i=\frac{20}{14+L} \text{ subject to maximum of } 1{\cdot}0$$

(*ii*) For main girders of double tracks $i = 0.72 \times \frac{20}{14+L}$ subject to a maximum of 0·72

(*iii*) For intermediate main girders of multiple track spans

$$i = 0.6 \frac{20}{14+L} \text{ subject to maximum of } 0.6$$

(*iv*) For outside main girders of multiple track spans with intermediate girders, the impact allowance shall be that specified in (*i*) and (*ii*) whichever applies

(*v*) For cross girders carrying two or more tracks

$$= 0.72 \times \frac{20}{14+L} \text{ subject to a maximum of } 0.72$$

(*vi*) For rails where ordinary fish plated joints are supported directly on transverse steel troughing or sleepers, the imfact factor shall be

$$\text{For Broad gauge } i = \frac{7.32}{B+5.49}$$

$$\text{Metre gauge } i = \frac{5.49}{B+4.27}$$

where B = the spacing of main girder in metres.

1·5. Wind Loads Unloaded Spans

For foot bridges and railway bridges the wind pressures are as per **IS-875**

(*i*) For unloaded spans and trestles the exposed area shall be considered as 1·5 times the horizontal projected area of the span or the trestle.

For plate girders, the area of the leeward girder shall be multiplied by the factors given below and should be added to the area of windward girder.

For the spacing of Leeward girder does not exceed half its depth $= 0$

For spacing exceeding $\frac{d}{2}$ and upto $d = 0.25$

For spacing exceeding d and upto $1.5d = 0.50$

For spacing exceeding $1.5d$ and upto $2d$ or more $= 1.00$.

1·5·1. Loaded Spans

For loaded spans the net exposed area shall be the total of

(*i*) 1·5 times that portion of the horizontal projected area of the span not covered by the moving load for all spans and types of bridges other than plate girder bridges.

For plate girder bridges, the area of leeward girder not covered by the moving load shall be multiplied by the factors given for unloaded spans and added to the area of windward girder.

(*ii*) The horizontal projected area of the moving load.

The area of the moving load shall be taken as from 600 mm above rail level to the top of the highest stock for which the bridge is designed.

For foot bridges the height of the moving load is to be taken as 2 m throughout the length of the span.

1·6. Lateral Loads

Lateral bracings of the loaded deck of railway spans shall be designed to resist, in addition to the wind and centrifugal loads, a lateral load due to racking forces of 600 kg/m treated as a moving load. This lateral load need not be taken into accouut when calculting stresses in chords and flanges of main girders.

Lateral bracings shall be designed for a lateral load due to wind and racking forces of 900 kg/m treated as a moving load in addition to the centrifugal load if any, for spans upto 20 m, where separate wind stresses need not be calculated.

1·7. Longitudinal Forces

Longitudinal loads arising from any one or more of the following causes.

(*i*) The tractive effort of the driven wheels of locomotives.

(*ii*) The tracking force resulting from the application of the brakes to all braked wheels.

(*iii*) Resistance to the movement of the bearings due to change of temperature.

These loads shall be considered as acting horizontally through the Knuckle pins or through the sliding bearings. For spans with sliding bearings the horizontal loads are divided equally between the two ends and for spans with roller bearings at one end, the entire horizontal load is considered to act through the fixed end.

Where the structure carries more than one track, longitudinal loads shall be considered to act simultaneously on all tracks.

1·8. Centrifugal Force

The horizontal load due to centrifugal force is assumed to act at a height of 1·83 m above the rail level.

$$C=\frac{wV^2}{127\,R}$$

C=horizontal effect in tonnes per m of span

w=equivalent distributed live load in tonnes/m

V=Maximum speed in km/hr

R=Radius of the curve in m

1·9. Bearings

Frictional resistance of the expansion bearings.

For roller bearings=0·03

Sliding bearings of steel on cast iron or steel bearings=0·25

Sliding bearings of steel or Ferro asbestos=0·20.

1·10. Loads for BM and SF

For broad gauge 1676 mm

Equivalent uniformly distributed loads on each track for calculating bending moments and shears are given in the table.

L metres	*Total load (tonnes) for bending moment*		*Total load (tonnes) for shear*		*Impact factor*
	M L.	*B.L.*	*M.L.*	*B.L.*	$\frac{20}{14+L}$
1·0	45·8	34·6	45·8	34·6	1·00
1·5	45·8	34·6	45·8	34·6	1·00
2·0	45·8	34·6	52·4	39·6	1·00
2·5	45·8	34·6	60·4	45·5	1·00
3·0	46·9	35·4	65·5	49·5	1·00
3·5	52·4	39·6	70·3	53·2	1·00
4·0	59·2	44·8	78·8	59·4	1·00
4·5	67·9	51·4	85·2	64·4	1·00
5·0	74·8	56·7	90·3	68·3	1·00
5·5	80·6	60·9	96·7	73·2	1·00
6·0	85·2	64·4	104·9	78·6	1·00
6·5	89·3	67·5	110·2	83·2	0·976
7·0	95·2	71·5	115·2	87·1	0·952
7·5	100·7	75·8	119·8	90·6	0·931
8·0	105·6	79·7	123·9	93·5	0·909
8·5	110.2	82·9	128·6	97·2	0·889
9·0	114·0	86·2	132·4	101·4	0·870
9·5	117·6	90·8	136·7	105·0	0·851
10	121·0	94·1	140·6	108·3	0·833
11	133·6	102·5	148·3	114·0	0·800
12	140·9	108·4	155·7	119·1	0·769
13	147·2	113·4	163·7	125·3	0·741
14	152·7	117·2	172·0	131·6	0·714
15	160·6	123·2	180·6	138·2	0·691
16	168·8	130·2	188·1	144·8	0·667
17	177·0	137·1	196·8	151·6	0·645
18	185·9	143·8	205·0	159·0	0·625
19	193·9	150·8	214·1	166·0	0·606
20	202·7	158·0	222·4	172·6	0·588
21	211·1	164·9	230·5	179·1	0·571
22	218·7	169·8	238·7	185·5	0·556
23	225·6	175·8	246·6	191·6	0·541
24	232·9	181·8	254·8	197·7	0·526
25	241·0	188·0	262·7	203·7	0·513
26	249·5	193·5	270·7	209·7	0·500
27	256·0	198·8	278·9	215·8	0·488
28	264·0	205·2	286·6	223·0	0·476
29	271·0	211·8	294·4	230·2	0·465
30	280·0	216·5	302·3	237·5	0·455

32	294·9	230·0	320·0	251·8	0·435
34	309·5	243·5	337·5	265·2	0·417
36	327·0	257·5	354·2	278·0	0·400
38	342·3	270·0	371·2	291·2	0·385
40	359·0	282·0	387·7	304·8	0·370
42	375·0	296·0	404·6	318·6	0·357
44	391·0	308·5	421·8	331·6	0·345
46	408·0	322·3	438·4	344·8	0·333
48	424·0	334·5	454·9	357·6	0·323
50	438·0	347·0	471·3	370·2	0·313
55	477·5	376·5	512·2	401·0	0·290
60	514·8	405·2	552·8	431·0	0·270
65	554·0	433·5	591·6	460·0	0·253
70	591·8	461·0	632·2	488·2	0·238
75	628·3	486·5	672·0	515·4	0·225
80	667·0	513·0	710·0	543·8	0·213
85	703·5	539·0	750·4	570·8	0·202
90	742·0	568·0	789·8	597·8	0·192
95	780·0	591·6	827·8	624·6	0·183
100	820·0	616·0	868·6	651·2	0·175
105	858·0	642·5	906·2	677·2	0·168
110	897·9	668·5	945·6	704·2	0·161
115	935·0	694·5	984·8	730·2	0·155
120	973·0	719·5	1025·3	756·4	0·149
125	1010·0	745·0	1072·4	782·4	0·144
130	1048·5	770·0	1113·3	880·4	0·139

For bending moment L is equal to the effective span in metres. For shear, L is equal to the loaded length in metres to give the maximum shear in the member under consideration.

1·11. Loads for BM and SF

For metre gauge 1000 mm

(*i*) Standard main line (ML)

(*ii*) Standard branch line (BL)

(*iii*) Standard C

Equivalent uniformly distributed loads in each track for calculating bending moments and shears are given in the table.

L *metres*	*Total load* (*tonnes*) *for bending moment*			*Total load* (*tonnes*) *for shear*			*Impact factor*
	ML standard	*BL standard*	*C loading*	*ML standard*	*BL standard*	*C. loading*	$\frac{20}{14+L}$
1·0	26·4	21·4	16·2	26·4	21·4	16·2	1·000
1·5	26·4	21·4	16·2	29·1	23·6	19·2	1·000
2·0	26·4	21·4	17·4	35·0	28·4	23·8	1·000
2·5	28·2	22·9	20·1	38·6	31·3	25·5	1·000
3·0	32·0	25·7	22·3	43·7	35·4	28·9	1·000
3·5	38·6	31·3	26·0	48·8	39·4	31·7	1·000
4·0	43·7	35·4	28·9	52·2	42·6	35·0	1·000
4·5	47·7	38·6	31·0	57·9	47·0	38·3	1·000
5·0	50·8	41·1	33·8	62·7	50·8	40·9	1·000
5·5	55·0	44·6	36·5	66·6	54·0	43·1	1·000
6·0	59·3	48·0	38·9	69·9	56·6	45·4	1·000
6.5	62·3	50·8	40·9	73·1	60·1	48·2	0·976
7·0	65·7	54·2	43·7	77·0	63·3	50·3	0·952
7·5	69·0	57·6	45·9	80·6	66·0	52·3	0·931
8·0	72·8	60·5	48·2	83·5	68·4	54·1	0·909
8·5	76·1	63·0	49·9	86·2	70·5	56·0	0·889
9·0	78·9	65·3	51·6	88·8	73·0	58·2	0·870
9·5	81·6	67·4	53·7	92·2	75·5	60·2	0·851
10·0	84·4	68·7	55·6	95·0	78·0	62·4	0·833
11·0	90·2	73·9	59·9	102·2	83·8	67·1	0·800
12·0	97·2	80·0	64·9	108·2	89·6	72·2	0·769
13·0	103·5	86·4	69·8	114·4	95·5	77·3	0·741
14·0	109·8	92·1	74·8	120·6	101·8	82·1	0·714
15·0	116·0	98·0	79·6	127·7	107·3	86·7	0·691
16·0	122·6	103·5	83·7	133·8	112·5	91·1	0·667
17·0	128·6	108·5	87·9	139·7	117·6	95·5	0·645
18·0	134·2	113·3	91·9	145·4	122·6	99·9	0·625
19·0	139·5	118·2	95·9	150·9	127·5	104·1	0·601
20·0	144·9	122·6	99·8	156·2	132·3	108·4	0·588
21·0	149·5	126·8	104·1	161·4	136·8	112·6	0·571
22·0	154·8	131·4	108·4	167·0	141·5	117·1	0·556
23·0	160·6	136·4	112·7	173·4	147·2	121·7	0·541
24·0	165·8	140·8	116·8	180·2	152·8	123·5	0·526
25·0	171·0	145·5	121·3	186·9	158·6	130·9	0·513
26·0	177·8	151·2	125·5	193·4	164·0	135·2	0·500
27·0	184·4	156·8	130·4	199·9	169·0	139·6	0·488

28·0	190·6	162·0	131·4	206·0	174·4	143·8	0·476
29·0	196·7	166·9	133·4	212:2	179·8	149·0	0·465
30·0	203·0	171·9	142·6	218·6	185·3	153·3	0·455
32·0	214·5	181·7	152.0	231·2	196·8	163·1	0·435
34·0	226·5	192·9	161·3	244·2	207·6	172·1	0·417
36·0	239·0	204·0	169·9	256·8	218·1	181·0	0·400
38·0	251·5	214·4	178·4	268·7	228·4	189·8	0·381
40·0	264·0	224·2	186·7	280·2	238·4	198·5	0·370
42·0	274·5	233·6	195·0	291·4	248·1	207·0	0·357
44·0	285·5	243·0	203·4	302·3	257·7	215·6	0·345
46·0	296·0	252·0	211·6	313·0	267·1	224·0	0·333
48·0	306·0	261·2	219·7	323·5	276·4	232·4	0·323
50·0	316·0	270·3	227·6	333·4	285·6	240·7	0·313
55·0	340·0	292·1	246·9	357·9	308·0	261·0	0·290
60·0	363·0	312·9	266·7	381·8	329·9	281·6	0·270
65·0	385·0	333·2	286·7	404·7	351·4	302·0	0·253
70·0	405·0	354·1	306·9	427·4	372·7	322·0	0·238
75·0	426·5	373·7	326·9	450·1	393·6	342·2	0·225
80·0	448·0	393·0	346·4	471·6	414·3	361·9	0·213
85·0	468·5	413·9	365·8	492·7	435·0	381·6	0·202
90·0	488·1	433·6	385·5	514·2	455·5	401·8	0·192
95·0	508·0	452·8	404·7	534·3	475·8	421·4	0·183
100·0	528·0	472·9	423·2	555·2	496·1	441·4	0·175
105·0	548·0	491·9	342·7	576·5	516·2	462·2	0·168
110·0	567·5	511·6	461·4	597·4	536·3	480·6	0·161
115·0	587·0	530·5	480·0	617·7	556·3	500·2	0·155
120·0	607·0	549·1	500·5	638-1	576·3	520·0	0·149
125·0	627·0	568·1	518·0	658·3	596·2	539·6	0·144
130·0	646·0	589·3	538·9	678·8	616·1	559·0	0·139

For bending moment, L is equal to the effective span in metres, for shear, L is equal to the loaded length in metres to give the maximum shear in the member under consideration.

2. Design of a Welded Plate Girder for a Railway Bridge (B.G Track)

2·1. Data

Span of the plate girder = 30 m

Broad gauge (1676 mm) Main line Single track

2·2. Dead Loads

Dead load per track (open floor) = 750 kg/m

Self weight of the girder = $20\,L+100$ kg/m = $20\times30+100$
= 700 kg/m

Total dead load = w = 1450 kg/m

2·2·1. Live Loads

Equivalent total live load for bending moment Calculation per track = 280 t

Total live load per girder = $W=\frac{280}{2}=140$ t

Equivalent total live load for shear force computations = 302·3 t

Total live load per girder for shear consideration

$$=\frac{302\cdot3}{2}=151\cdot15 \text{ t}$$

2·3. Impact Factor on Live Loads

Impact factor on live loads $=\frac{20}{14+L}=\frac{20}{14+30}=0\cdot455$

2·4. Bending Moments

B.M. due to dead load $=\frac{wL^2}{8}=\frac{1450\times30\times30}{8\times1000}$ = 163 mt

B.M. due to live load $=\frac{WL}{8}=\frac{140\times30}{8}=526$ mt

BM due to impact on Live loads = $0\cdot455\times526=235$ mt

Total BM = M = 924 mt

2·5. Shear Forces

SF due to dead load $=\frac{1450\times30}{2\times1000}=21\cdot75$ t

SF due to live load $= \frac{151{\cdot}15}{2} = 75{\cdot}57$ t

SF due to impact on live load $= 75{\cdot}57 \times 0{\cdot}455 = 34{\cdot}5$ t

Total shear force $= S = 131{\cdot}82$ t

2·6. Proportioning Trial Section Web Plate

Aproximate depth of girder

$$= \frac{1}{8} \text{ to } \frac{1}{10} \text{ of span} = \frac{3000}{100} = 300 \text{ cm}$$

$$\text{Economical depth} = 5\sqrt[3]{\frac{M}{f_b}} = 5\sqrt[3]{\frac{92400000}{1575}} = 194 \text{ cm}$$

Web depth on shear consideration assuming 12 mm thick plate (since t=10 mm for Railway bridges)

$$d = \frac{S}{q_s \times 1{\cdot}2} = \frac{131820}{945 \times 1{\cdot}2} = 116 \text{ cm}$$

Try 160 cm × 1·2 cm

2·6·1. Flange Plates

Approximate flange area required

$$A_f = \frac{M}{f_b\, d} - \frac{A_w}{6} = \frac{92400000}{1575 \times 160} - \frac{1{\cdot}2 \times 160}{6}$$

$$= 368 - 32 = 336 \text{ cm}^2$$

$$\text{Flange width} = B = \frac{L}{40} \text{ to } \frac{L}{45} = \frac{3000}{40} \text{ to } \frac{3000}{45}$$

$$= 75 \text{ cm to } 67 \text{ cm}$$

Use $B = 70$ cm

$$\text{The thickness of plate required} = \frac{A_f}{B} = \frac{336}{70} = 48 \text{ cm}$$

Use flange plate of 70 cm × 5 cm

2·6·2. Cbeck for Maximum Stresses

$$\text{Moment of intertia} = I = \frac{1}{12}\, t_w\, d^3 + 2 \times A_f \times (d')^2$$

$$= \frac{1}{12} \times 1{\cdot}2 \times 160^3 + 2 \times 350 \times 82{\cdot}5^2 = 411{,}000 + 4{,}770{,}000$$

$$= 5181{,}000 \text{ cm}^4$$

2·6·3. Bending Stress

$$\text{Max. bending stress } f_b = \frac{M}{I}\, y = \frac{92{,}400{,}000}{5{,}181{,}000} \times 85$$

$$= 1520 \text{ kg/cm}^2$$

2·6·4. Shear Stress

$$\text{Max. shear stress} = \frac{131820}{160 \times 1{\cdot}2} = 685 \text{ kg/cm}^2$$

The trial section is therefore safe.

2·6·5. Self Weight Check

Weight of web plate/m $= 1{\cdot}60 \times 1 \times \dfrac{1{\cdot}2}{100} \times 7850 = 150{\cdot}4$ kg/m

Weight of flange plates/m length

$= 2 \times \dfrac{70}{100} \times \dfrac{50}{100} \times 1 \times 7850 = 550$ kg/m

Add 30% of web weight for welding $= \dfrac{150{\cdot}4 \times 30}{100} = 45{\cdot}12$ kg/m

Total $= 700{\cdot}4 + 45{\cdot}12 = 745{\cdot}52$ kg/m close to 700 kg/m assumed.

2·7. Connection Between the Flange and Web

Maximum shear at the junction of web and flange

$q = \dfrac{Fay}{I}$

$F = 131820$ kg

$a = 70 \times 5 = 350$ cm^2

$I = 5{,}181{,}000$ cm^4

$y = 82{\cdot}5$

$q = \dfrac{131820 \times 350 \times 82{\cdot}5}{5{,}181{,}000} = 735$ kg/cm

Size of the weld required assuming continuous weld on either side.

Strength of weld of size $S = 2 \times 0{\cdot}7 \times S \times 1025 = 1450\ S$, kg/cm

$1450\ S = 735$

$S = \dfrac{735}{1450} = {\cdot}52$ cm

Use 6 mm fillet weld continuous on either side

2·8 Stiffners

$\dfrac{d}{t} = \dfrac{160}{1{\cdot}2} = 133 < 200$, no longitudinal stiffners are to be provided.

Since $\dfrac{d}{t} > 85$ Intermediate stiffeners are to be provided

2·8·1. Intermediate Stiffners

Spacing of the stiffners

$= \dfrac{d}{3}$ or $1{\cdot}5\ d = \dfrac{160}{3}$ or $1{\cdot}5 \times 160 = 53$ cm or 240 cm

Use 150 cm spacing

2·8·2. Panel Dimensions

Greatest unsupported panel dimension of the web

160 cm $< 270 \times tw = 270 \times 1{\cdot}2 = 324$ cm

The intermediate stiffeners are designed to have a minimum moment of inertia

$$I=\frac{1{\cdot}5\, d'^3\, tw^3}{C^2}=\frac{1{\cdot}5\times 160^3\times 1{\cdot}2^3}{150\times 150}=471 \text{ cm}^4$$

Use 10 mm plate

Maximum width of plate $\not> 16\,t \not> 16$ cm

Use a plate of 1 cm $\times$ 12 cm

$$I=\frac{1}{3}\times 12^3\times 1=576 \text{ cm}^4>471 \text{ cm}^4$$

2·8·3. Connections to Vertical Stiffner to Web

Shear on welds connecting stiffener to web

$$=\frac{t^2_w}{2h} \text{ tonnes/cm}=\frac{1{\cdot}2^2}{2\times 12}={\cdot}06 \text{ t/cm}=60 \text{ kg/cm}$$

$$\text{Size of the weld}=\frac{60}{{\cdot}7\times 1025}={\cdot}0834 \text{ cm}$$

Use 5 mm minimum size intermittent fillet weld (upto 19 mm thick plate)

Effective length of weld $\not< 10\,t=10\times 1{\cdot}2=12$ cm

16 cm length 5 mm size fillet weld alternately on either side

2·8·4. End Bearing Stiffners

Maximum shear force=131·82 t

The end stiffener is designed as a column.

Bearing stiffener shouid be extended as nearly as practicable to to the outer edges of the flanges

$$\frac{b}{t}\not> 12 \text{ or } t=\text{the thickness of the plate is} \not<\frac{b}{12}$$

Use $b=30$ cm

$$t=\frac{30}{12}=2{\cdot}5 \text{ cm}$$

Use 30 cm $\times$ 2·5 cm plate

$$\text{Bearing area required}=\frac{131{\cdot}82}{1{\cdot}890}=69{\cdot}5 \text{ cm}^2$$

If two plates are used total area provided

$=30\times 2{\cdot}5\times 2=150 \text{ cm}^2>69{\cdot}5 \text{ cm}^2$

The length of web plate which acts along with the stiffener plates in bearing the reaction$=20\,t=20\times 1{\cdot}2=24$ cm

Moment of inertia I

$$=\frac{1}{12}\times 2{\cdot}5\times 61{\cdot}2^3+\frac{2\times 24\times 1{\cdot}2^3}{12}=47{,}900+6{\cdot}9=47906{\cdot}9 \text{ cm}^4$$

$\text{Area}=61{\cdot}2\times 2{\cdot}5+48\times 1{\cdot}2=151+57{\cdot}8=208{\cdot}8 \text{ cm}^2$

$$r_{min}=\sqrt{\frac{47906{\cdot}0}{208{\cdot}8}}=15{\cdot}1 \text{ cm}$$

Effective length of the stiffener$=160\times 0{\cdot}7=112$ cm

$l/r_{min}=\dfrac{112}{15{\cdot}1}=7{\cdot}4$

Allowable stress $=1247$ kg/cm²

Area required to resist reaction $=\dfrac{131820}{1247}=104\ \text{cm}^2<208{\cdot}8\ \text{cm}^2$

2·8·5. Weld Connection

Length available for welding using alternate intermittent weld

$=2\,(160-4)=312$ cm

Required strength of weld $=\dfrac{131820}{312}=416$ kg/cm

Size of the weld $=\dfrac{416}{\cdot 7\times 1025}=\cdot 58$

Use 6 mm fillet intermittent weld

Length of weld ≮ 10 t ≮ $12\times 1{\cdot}2=14{\cdot}4$ cm

Use 16 cm long 6 mm size fillet welds alternately.

2·9. Spacing of Girder

$$\left.\begin{array}{l} \dfrac{L}{20}=\dfrac{300}{20}=15\ \text{m} \\ \not< \dfrac{L}{15}=\dfrac{300}{15}=200\ \text{m} \\ \qquad 2{\cdot}25\ \text{m minimum} \end{array}\right\}\ \text{for} > 20\ \text{m spans}$$

2·9·1. Lateral Bracing

For resisting wind, racking and centrifugal forces Lateral, bracing is provided. End cross frames are also provided for spans greater than 20 m

2·9·2. Wind Loads

Wind load $=150$ kg/m²

Spacing between two girders $=1{\cdot}676$ m

Spacing to depth of girder $=\dfrac{1{\cdot}676}{1{\cdot}7}=0{\cdot}985$

Coefficient for wind load on Leeward girder $=0{\cdot}25$

Wind load on the wind ward girder $=150\times 1{\cdot}7\times 30=7660$ kg

Add wind load on leeward girder $=0{\cdot}25\times 7660=1915$ kg

Total wind load $=7660+1915=9575$ kg

Lateral load due to racking forces $=600$ kg/m

Total racking force $=600\times 30=18{,}000$ kg

Total lateral load $=18000+9575=27{,}575$ kg

Maximum tension in the diagonal due to lateral load

$$=\frac{14}{15}\times\sec\theta\times\tfrac{1}{2}\times 27575=\frac{14}{15}\times\frac{\sqrt{2^2+1{\cdot}676^2}}{2}\times\tfrac{1}{2}\times 27575$$

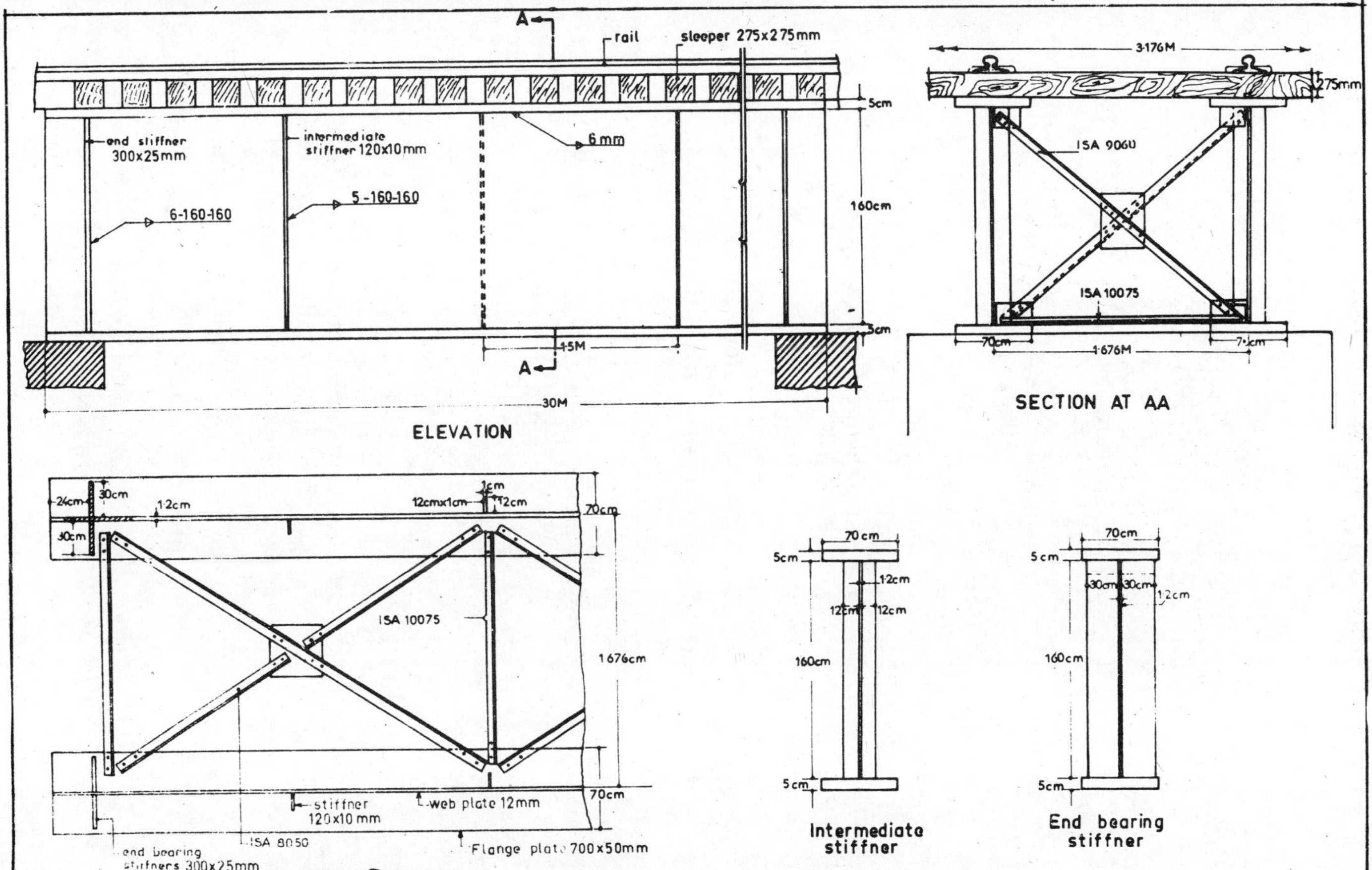

RAILWAY BRIDGE - BG TRACK - 30M SPAN

$=\frac{14}{15}\times\frac{\sqrt{4+2\cdot8}}{2}\times\frac{1}{2}\times27575=\frac{14}{15}\times\frac{2\cdot6}{2}\times\frac{1}{2}\times27575$

$=16{,}800$ kg

Area required $=\frac{16800}{1500}=11\cdot2$ cm^2

Use ISA – 8050 with 10 mm thickness
Area provided $=12\cdot02$ cm^2

Maximum compressiveForce in $U_1\,L_1=\frac{27575}{2}$ kg $=13787\cdot5$ kg

Length of member $=1\cdot676$ m
Effective length $=1\cdot676\times0\cdot85=1\cdot43$ m
Try an angle – ISA 10075 with 10 m thick
Area $=16\cdot50$ cm^2
$r_{min}=2\cdot16$ cm

$l/r_{min}=\frac{143}{2\cdot16}=66$

Allowable stress $=1097$ kg/cm^2
Safe load that the member can carry $=16\cdot5\times1097$
$=18{,}100$ kg >13789 kg

2·10. Cross Frame

Lateral load to be resisted by one frame $=13787\cdot5$ kg

Tension in the diagonal $=\frac{13787\cdot5}{\text{Cos }\theta}=\frac{13787\cdot5}{1\cdot676}\times2\cdot31$

$=19100$ kg

Area required $=\frac{19100}{1500}=12\cdot75$ cm^2

Use – ISA
Use – ISA 9060, 10 mm thick
Area provided $=14\cdot01$ cm^2

PART IV

DESIGN OF STEEL WATER TANKS

1. Steel Tanks—General

1·1. General

Steel tanks are used to store water, gas and other liquids. These tanks are often used in railway yards to store water for locomotives and in industrial water supply schemes.

1·2. Types

(*i*) Riveted water tanks
 (*a*) Square or (*b*) Rectangular
(*ii*) Pressed steel water tanks
 (*a*) Square or (*b*) Rectangular
(*iii*) Circular water tanks
 (*a*) with hemisperical bottom
 (*b*) with conical bottom
 (*c*) stand pipes—cylindrical tanks with flat bottom

1·3. Riveted Water Tanks (Tank Plates)

The tank plates are generally 1·25 m wide. The minimum thickness of plates used in tanks is 6 mm. The bottom plates at the edges are turned up at the ends to facilitate proper jointing with the side plates.

1·4. Joints (T–Cover)

All the joints are riveted. when ordinary steel plates are used. The joints for bottom plates are covered with Tee—or angle sections. These are turned up vertically on the side plates of the tank and these will act as vertical stiffners to the side plates. The horizontal and vertical joints are covered with steel flats.

1·5. Stays

Usually steel flats are used as stays. These are used at mid height of the tank and are connected to vertical stiffners of the side plates. These are provided in both the directions for holding the side plates of the tank. These are designed to hold the overhanging portion of the tank beyond the extreme beams on which the tank is supported.

1·6. Supporting Beams

The top tier of the beams are placed below the T-cover joints so that, these joints transfer the load to the beams below. The lower tier of the beams rest directly on columns.

1·7. Columns

Columns should be placed such that they offer maximum resistance against overturning. But however, increase in spacing of columns increases bending moment in the bottom tier beams. Hence a compromise has to be made between the spacing of bottom tier beams and the spacing of columns by giving a suitable overhang to the beams supporting the tank, to obtain maximum economy. The effective length of column is taken as centre to centre of bracing.

1·8. Bracings

In order to take the wind pressure acting on the tank, the side diagonal bracings are required. These bracings also provide lateral restraint to the main columns. Horizontal bracings are provided at the ends of diagonal side bracings. These bracings in turn are to be cross braced to make the horizontal frame stable, against the wind blowing diagonally.

1.9. Foundations

Columns transmit the load to the concrete foundation through base plate. The essential requirement is that the centre of gravity of the column section should coincide with the centre of gravity of base plate.

1·10. Pressed Steel Tanks General

These are fabricated with standard size pressed plates of 1·25 m ×1·25 m using bolting. Steel plates are pressed to shape with flanges at the edges. These tanks are used upto a maximum depth of **5 m.** Plastic compound is used to prevent leakage in the joints.

1·10·1. Plates

Plates are available in standare sizes of 1·25 m×1·25 m. The thickness of plates vary from 3, 5, 6 and 8 mm. Since the plate sizes are fixed, the length width and depth of tanks are multiples of 1·25 m. These plates may have flanges on the external side or flanges inside of the tank. Some tanks may have bottom plates with flanges internal and side plates with flanges external.

1·10·2. Capacity of Tank

Depth in metres	*Capacity in litres*
1·25	1950 to 95,550
2·50	7800 to 191,100
3·75	52,650 to 988,650
5·00	73,200 to 1,318,200

1·10·3. Thickness of Plates

The thickness of plates depend on the water pressure therefore it would depend upon the location of the plate.

Depth of tank *m*	*Location of plate*	*Thickness of plates in mm*
1·25	bottom plates	5
	side plates	5

2·50	top plates	5
	second tier of plates	6
	bottom plates	6
3·75	top plates	5
	second tier plates	6
	third tier plates	8
	bottom tier plates	8
5·00	top plates	5
	second tier	6
	third tier	6
	fourth tier	8
	bottom plates	8

1·10·4. Weight of Plates

Thickness mm	*Weight in kg*
3	52
5	86
6	112
8	138

1·10·5. Stays

Since the bolted joints cannot take any bending moment the sides of the tank are to be supported by stays. These stays are usually made up of mild steel bars or rolled sections. These are either directly bolted to flanges or through M.S cleats.

1·11. Cylindrical Tanks General

Minimum thickness of the plates in cylindrical portion is 6 mm. The thickness of the plates in the bottom shall not be less than thickness of the plate in the lower course of the cylindrical portion.

1·11·1. Joints

Radial seams in the bottom shall be Lap joints. Lap joints are also used for horizontal seams. For vertical joints for the plates from 6 to 8 mm, double-riveted lap joints and for plates from 10 mm to 12 mm triple riveted joints are used. 70 to 75% of efficiency is expected from the joints.

1·11·2. Rivet Size

Thickness of plate mm	*Diameter of rivet* mm
6 mm	16 mm
8 mm	20 mm
10 mm	22 mm
12 mm	22 mm

1·11·3. Stresses In Tanks With Hemispherical Bottom

p=the pressure at any depth due to water
d=diameter of the cylinderical portion
t=thickness of the plate

Hoop stress in the cylindrical plate $= \dfrac{pd}{2t} = \dfrac{whd}{2t}$ kg/m²

STEEL WATER TANK

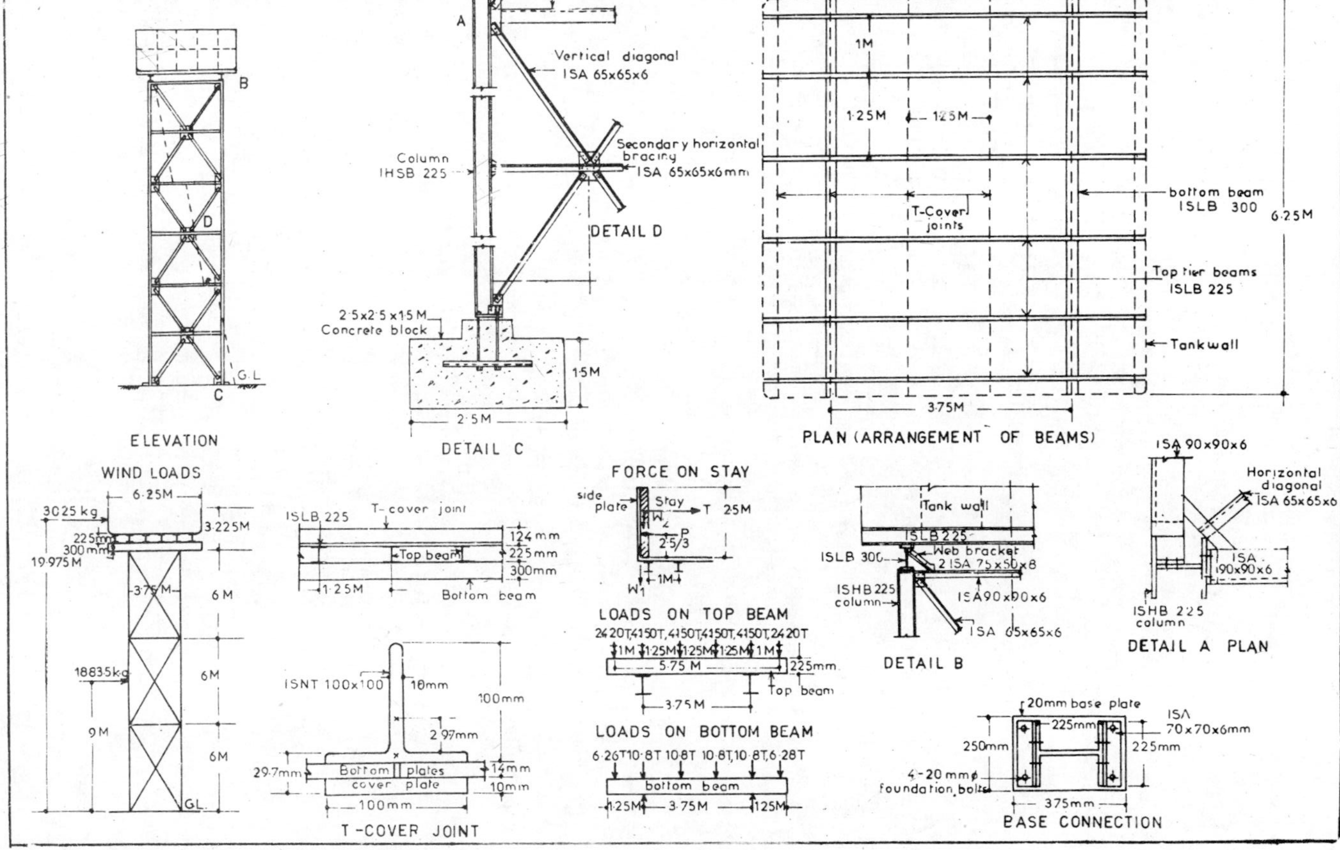

Maximum radial stress and circumferential stress in the bottom radial and circumferential joints.

$$=\frac{wHR}{2t} \text{ or } =\frac{wHd}{4t} \text{ kg/m}^2$$

Maximum stresses in circumferential joints in the cylindrical

$$=\frac{whR}{2t} \text{ kg/m}^2$$

1·11. Tanks with Conical Bottom

Maximum stress on radial joints $= \frac{whd}{4t}$ Cosec θ kg/m²

Maximum stress on circumferential joint$=\frac{whd \operatorname{cosec} \theta}{2t}$ kg/m²

1·12·1. Connection between the cylindrical portion and the conical portion

Maximum radial stress$=\frac{w\,h\,d}{4\,t}$ coses θ kg/m²

Maximum radial tension$=\frac{w\,h\,d}{4}$ cosec θ kg

Compressive force on the ring$=\frac{w\,h\,d}{4} \operatorname{cosec} \theta \times \frac{d \cos \theta}{2}$

$$=\frac{w\,h\,d^2 \cot \theta}{8} \text{ kg}$$

1·12·2. The Ring Design

The ring is designed as a street with slenderness ratio

$$=\frac{d}{2\sqrt{3}\,r}$$

where r=radius of gyration of the ring about vertical axis.

For tanks with hemispherlcal bottoms hoop compression is zero.

1·12·3. Circular Girder

A circular girder is used to transfer the load of the tank and the contents to the colums. It is supported on four or more columns

Bending moments and torsion in circular girders

W=Total load on the girder
R=Radius of the girder

Number of Columns	*BM at support*	*BM at centre*	*Maximum torsion*	*Position af max. torsion*
4	0 0342 WR	0 0176 WR	0 0053 WR	19°·12′
6	0·0148 WR	0 0075 WR	0 0015 WR	12°·44′
8	0·0083 WR	0 00416 WR	0·006 WR	9°·33′
10	0 0054 WR	0 0023 WR	0·0003 WR	7°·30′
12	0·00375 WR	0 00142 WR	0·00017 WR	6°·15′

2. Elevated Square Steel Tank

2·1. Data

Capacity = 90 cu m; Free board = 10 cm

Height of the tank from *GL* to the bottom of the lower tier beams = 18 m

Materials available: Plates of width 1·25 m; Rolled and angle sections of all sizes.

Bearing capacity of soil = 15 t/m²; Bearings on concrete = 40 kg/cm².

2·2. Relevant Codes

IS—804—1958; *IS*—800—1962; *IS*—875—1964.

2·2·1. Allowable Stresses

In plates in bending = 1650 kg/cm²; Rivets in shear = 945 kg/cm²
Trial Dimensions.

2·3. Trial Dimensions

Capacity = 90 cu m; Depth of tank = 2·5 m; Depth of water = 2·4 m

$$\text{Area in plan of the tank} = \frac{90}{2{\cdot}4} = 37{\cdot}5 \text{ m}^2$$

Side of the tank = 6·15 m
Using 5 plates of width 1·25 m each side of the tank = 6·25 m
Use 6·25 m × 6·25 m × 2·5 m
Capacity provided = 6·25 × 6·25 × 2·4 = 94 cu m > 90 cu m.
Consider 1 m wide plate spanning between two T-cover joints

2·4. Tank portion bottom plates

Water load = 2·5 × 1000 = 2500 kg/m
Self weight (assuming 15 mm thick) = 120 kg/m
Total = 2620 kg/m

$$\text{Maximum bending moment } M = \frac{wL^2}{8} = \frac{2620 \times 1{\cdot}25^2}{8}$$

$$= 511 \text{ m kg} = 51100 \text{ cm kg}$$

$$M = fZ$$

$$Z \text{ required} = \frac{51100}{1650} = 31 \text{ cm}^3$$

If t = thickness of the plate

$$Z = \tfrac{1}{6}\, b\, t^2$$

$$t^2 = \frac{31 \times 6}{100}$$

$$t = \sqrt{1{\cdot}86} = 1{\cdot}36 \text{ cm}$$

Use 14 mm thick plate

2·4·1. T-Cover Joint

The joint is continuous over the top tier beams

Span $=1{\cdot}25$ m

Load on the joint $=1{\cdot}25\times1{\cdot}25\times2620=4100$ kg

Self weight of the T-cover joint $=50$ kg

Total load $W=4150$ kg

$$\text{Bending moment}=\frac{WL}{10}=\frac{4150\times1{\cdot}25}{10}=519 \text{ m kg}$$

$$=51900 \text{ cm kg}$$

2·4·2. Trial Section for the Joint

Use ISNT 100×100 T-section

and 10 cm $\times$ 1 cm cover plate

Area of T-section $=19{\cdot}1$ cm^2

$I_{xx}=173{\cdot}8$ cm^4

C.G of T-section $=2{\cdot}79$ cm from base

$$\text{C.G of the arrangement}=\frac{10\times2{\cdot}4\times1{\cdot}2+19{\cdot}10\,(2{\cdot}4+2{\cdot}79)}{10\times2{\cdot}4+19{\cdot}10}$$

$$=2{\cdot}97 \text{ cm}$$

$$I=\text{moment of inertia}=\frac{1}{12}\times10\times2{\cdot}4^3+2{\cdot}4\times10$$

$$\times\left(2{\cdot}97-\frac{2{\cdot}4}{2}\right)^2+173{\cdot}8+19{\cdot}10\,(2{\cdot}79-{\cdot}57)^2$$

$$=11{\cdot}5+75{\cdot}19+173{\cdot}8+94{\cdot}14=354{\cdot}63 \text{ cm}^4$$

$$Z=\frac{I}{y}=\frac{354{\cdot}63}{12{\cdot}4-2{\cdot}97}=\frac{354{\cdot}63}{9{\cdot}43}=37{\cdot}5 \text{ cm}^3$$

2·4·3. Stress in the Joint

$$\text{Stress in the joint}=\frac{M}{Z}=\frac{51900}{37{\cdot}5}=1380 \text{ kg/cm}^2<1575 \text{ kg/cm}^2$$

2·4·4. Rivet Design

$$\text{Shear force}=\frac{4150}{2} \text{ kg}$$

Using 16 mm ϕ rivets strength of rivet in single shear

$$=\frac{\pi\times1{\cdot}75^2}{4}\times945=2270 \text{ kg}$$

$$\text{Horizontal shear per cm length of the joint}=\frac{SA\bar{y}}{I}$$

$$S=\frac{4150}{2}=2075 \text{ kg}$$

$$A=2{\cdot}4\times10=24 \text{ cm}^2$$

$$\bar{y}=\left(2{\cdot}97-\frac{2{\cdot}4}{2}\right)=1{\cdot}77 \text{ cm}$$

$$SA\bar{y}=\frac{2075\times25\times1{\cdot}77}{354{\cdot}63}=249 \text{ kg}$$

Using 2 rivets in a row

$$\text{pitch of rivets} = \frac{2 \times 2270}{249} = \frac{4540}{249} = 18{\cdot}2 \text{ cm}$$

Maximum pitch allowed = 10 times the thickness of the thinnest plate

∴ Maximum pitch = $10 \times 1 = 10$ cm

Use 16 mm ϕ at 10 cm c/c.

2·5. Vertical Sides and Stiffners

Use 14 mm plates and the joints

Similar to the T-cover joints with 1 cm thick cover plate. T-sections will act as vertical stiffners. Assuming that stays are provided at mid depth of the tank in both directions connecting T-joints, the bending moment in the vertical stiffner at the level of stay

$$= \frac{wh^3_1}{6} \times l$$

$$\text{Maximum bending moment} = \frac{1000 \times 1{\cdot}25^3}{6} \times 1{\cdot}25 = 407 \text{ m kg}$$

$$= 40700 \text{ cm kg}$$

which is less than the bending moment on the bottom T-cover joint. Therefore, the same T-cover will be safe for vertical joint as well.

2·5·1. Stay

Stays are used in both directions. They connect the vertical T-joints at mid height of the tank. The vertical plates are assumed to be hinged to bottom plates at the junction.

Pressure on one vertical joint

$$P = \frac{wh^2}{2} \times \text{width of plate} = 1000 \times \frac{2{\cdot}5^2}{2} \times 1{\cdot}25 = 3905 \text{ kg}$$

Self weight of plate $= 1{\cdot}25 \times 2{\cdot}5 \times {\cdot}14 \times 7850$

$W_1 = 345$ kg

Weight of water in the overhung portion

$W_2 = 0{\cdot}25 \times 1{\cdot}25 \times 2{\cdot}5 \times 1000 = 775$ kg

Taking moments of all the forces about the hinge

$$T \times 1{\cdot}25 = P \times \frac{2{\cdot}5}{3} + W_1 \times 0{\cdot}257 + W_2 \times \frac{0{\cdot}25}{2}$$

$$T \times 1{\cdot}25 = 3905 \times \frac{2{\cdot}5}{3} + 345 \times {\cdot}257 + 775 \times \frac{0{\cdot}25}{2}$$

$$= 3250 + 86 + 96{\cdot}8 = 3432{\cdot}8$$

$$T = \frac{3432{\cdot}8}{1{\cdot}25} = 2750 \text{ kg}$$

Tension in the stay = 2750 kg

Taking allowable stress in tension for plates-flats = 1500 kg/cm²

$$\text{Area required} = \frac{2750}{1500} = 1{\cdot}83 \text{ cm}^2$$

Use 6 cm$\times$1 cm flats

Since stays run in both directions at same level and rest on those running in the perpendicular direction and therefore bending stresses due to self weight is releived since the unsupported length is 1·25 m only.

2·5·2. Connection between side Stiffner and the Stay

Maximum pull in the stay=2750 kg

Strength in single shear 16 mm ϕ rivet=2270 kg

$$\text{Number of rivets required}=\frac{2750}{2270}=1{\cdot}2$$

Use 2 rivets of 16 mm ϕ

2 6. Supporting Beams (Upper Tier Beams)

6 beams are used in the top tier.

Loads on intermediate beam from T-cover joint=4150 kg

$$\text{Load on the end beam}=\frac{4150}{2}+W_1=2075+345=2420 \text{ kg}$$

Self weight of the beam (assume)=50 kg/m

Bending moment due to concentrated loads

Reaction R_1=4150+4150+2420=10720 kg=10·72 t

Bending moment at support=2·420$\times$1=2·420 mt

$$\text{Bending moment at centre}=R_1\times\frac{3{\cdot}75}{2}-2{\cdot}42\times2{\cdot}875-4{\cdot}15\times1{\cdot}875-4{\cdot}15\times{\cdot}625$$

$$\text{Bending moment at centre}=10{\cdot}72\times\frac{3{\cdot}75}{2}-2{\cdot}42\times2{\cdot}875-4{\cdot}15(1{\cdot}875+{\cdot}625)$$

$$=20{\cdot}20-6{\cdot}98-10{\cdot}4=20{\cdot}20-17{\cdot}38=2{\cdot}82 \text{ mt}$$

Bending moment due to self weight at support

$$=\frac{Wh^2}{2}=\frac{50\times1{\cdot}25^2}{2}=39{\cdot}2 \text{ m kg.}$$

$$\text{at centre}=\frac{50\times6{\cdot}25}{2}\times\frac{3{\cdot}75}{2}-\frac{50\times6{\cdot}25}{2}\times\frac{6{\cdot}25}{4}$$

$$=48{\cdot}8 \text{ m kg}$$

Total moment at support=2420+39·2=2459·2 m kg

Total moment at centre =2820+48·8=2868·8 m kg

Maximum bending moment=286880 cm kg

$$Z \text{ required}=\frac{286880}{1650}=174{\cdot}5 \text{ cm}^3$$

Use ISLB 225 weighing 23·5 kg/m

Z provided=22?·4 cm^3

2·6·1. Shear Stress in the Beam

Maximum shear force=10·72−4·15=6·57 t=6570 kg

$$\text{Shear stress}=\frac{6570}{{\cdot}58}\times(22{\cdot}5-1{\cdot}72)=\frac{6570}{{\cdot}58\times20{\cdot}76}=545 \text{ kg/cm}^2$$

$$<945 \text{ kg/cm}^2$$

2·7. Bottom Tier Beams

Reaction from the intermediate top tier beam

$$=10720+\frac{23·5\times6·25}{2}=10720+73·5=10793·5 \text{ kg Say } 10·8 \text{ t}$$

Reaction from the edge top tier beam $=\frac{10·8}{2}$

$$+\frac{\text{weight of sidewall plates}}{2}=\frac{10·8}{2}+0·345\times\frac{6·25}{1·25\times2}$$

$$=5·4+·86=6·26 \text{ t.}$$

Bending moment due to concentrated loads

Maximum bending moment at support $6·26\times1=6·26$ mt

$$\text{BM at centre}=R_1\times\frac{3·75}{2}-6·26\times2·875-10·8\times1·875$$

$$-10·8\times0·625$$

$$=27·86\times\frac{3·75}{2}-6·26\times2·875-10·8\times(1·875+0·625)$$

$$=52·1-18-27=7·1 \text{ mt}$$

Bending moment due to self *wt* (assume 100 kg/m as selt *wt*.)

$$\text{BM at support}=100\times\frac{1·25}{2}\times1·25=78·5 \text{ m kg}$$

$$\text{Bending moment at centre}=\frac{100\times6·25}{2}\times\frac{3·75}{2}-100\times\frac{6·25}{2}$$

$$\times78·5 \text{ m kg}=97·6 \text{ m kg}$$

Total BM at centre $=710000+9760=719760$ cm kg

$$Z \text{ required}=\frac{719760}{1650}=436 \text{ cm}^3$$

Use ISLB–300 (Z provided $=488·7$ cm^3) (37·7 kg/m)

Max. reaction $=27·86$ t

Shear force $=27·86-10·8=17·06$ t

$$\text{Shear stress}=\frac{17060}{0·67\times28·12}=905 \text{ kg/cm}^2<945 \text{ kg/cm}^2$$

2·8. Columns Axial Load

Axial load on each column due to water

Self *wt* of tank + top tier and bottom tier beams $=27·86$ t

Load on the column due to self *wt* of beam (37·7 kg/m)

$$=\frac{37·7\times6·25}{2}=118 \text{ kg}$$

Total load on column $=27860+118=27978$ kg

Self *wt* of column (assume $=3022$ kg)

Total $=31000$ kg $=31$ t

2·8·1. Wind Load Normal To Face

Wind pressure $=150$ kg/m^2

Wind load on the tank and supporting beams
$=6 \cdot 25\times3\cdot025\times150=2850$ kg
This is assumed to act at mid height of tank i.e. 1·25 m from top
The point of application of this force above GL $=18+\cdot3+\cdot225$
$+1\cdot25=19\cdot775$ m
Wind load on columns (assume 20 cm wide columns)
$=2\times0\cdot2\times18\times150=1080$ kg

Wind load on horizontal bracings (assume 15 cm exposed width) at three levels $=3\cdot75\times0\cdot15\times3\times150=252\cdot5$ kg

Wind load on diagonal bracings (assuming 10 cm exposed width) at three levels $=6\times\sqrt{6^2+3\cdot75^2}\times0\cdot10\times150$

$=6\times7\cdot45\times0\cdot1\times150=671$ kg
Total wind load on the columns and bracings $=1080+252\cdot5$
$+671=2303\cdot5$ kg

The point of application of this force $=\frac{18}{2}=9$ m

2·8·2. Wind Load Moment

Wind load moment at the base of column,
$M_w=2850\times19\cdot775+2003\cdot5\times9=56400+18031\cdot5$
$=74432$ m kg
Force in each column due to wind moment

$$=\frac{M_w\,r}{\Sigma r^2}=\frac{74432\times\frac{3\cdot75}{2}}{4\times\left(\frac{3\cdot75}{2}\right)^2}=\frac{139500}{14\cdot15}=9860 \text{ kg}$$

2·8·3. Wind Load accross the Diagonal

Wind loads on the tank acting on the diagonal $=1\cdot41\times6\cdot25$
$\times3\cdot025\times150\times0\cdot9=3600$ kg
Point of application of this load $=19\cdot775$ m
Wind load on columns (Assuming 20 cm $\times$ 20 cm)
$=3\times\cdot2\times18\times150=1625$ kg

Wind load on diagonal bracing $=12\times\frac{7\cdot45}{2}\times0\cdot1\times150=955$ kg

Wind load on horizontal bracing $=6\times\frac{3\cdot75}{2}\times0\cdot15\times150=360$ kg

Total wind load on staging $=1625+955+360=2940$

2·8·4. Wind Load Moment

Point of application of this load $=9$ m
Wind load moment at the base of the column
$=3600\times19\cdot775+2940\times9=72000+26460=98{,}460$ m kg
Force in each column due to wind moment

$$=\frac{Mr}{\Sigma r^2}=\frac{98{,}460\times\frac{3\cdot75}{2}\sqrt{2}}{2\times\left(\frac{3\cdot75\sqrt{2}}{2}\right)^2}$$

$$=\frac{98,460\times 2\cdot 65}{2\times 7\cdot 04}=18,500 \text{ kg}$$

2·9. Design Columns

When tank is full

Load on the leeward column $=31+18\cdot5=49\cdot5$ t

Load on the windward column $=31-18\cdot5=12\cdot5$ t

When tank is empty

Load on the leeward column $=31-\frac{1}{4}$ of water load $+18\cdot5$

$$=31-\frac{94}{4}+18\cdot5=31-23\cdot5+18\cdot5=26 \text{ t}$$

Load on the windward column $=7\cdot5-18\cdot5=-11$ t

Effective length of column $=600$ cm

Maximum design load $=49\cdot5$ (comp)

Try ISHB 200

$$r_{vv}=4\cdot42 \text{ cm}$$

$$\frac{l}{r}=\frac{600}{4\cdot42}=136$$

Allowable stress (for $\frac{l}{r}=136$) $=531$ kg/cm^2

This can be increased by $33\frac{1}{3}\%$ since wind loads are considered

Increased allowable stress $=710$ kg/cm^2

Area of section $=50\cdot94$ cm^2

Permissible load on the column $=50\cdot94\times 710=36\cdot2$ t $<49\cdot5$ t

Hence unsafe

Try ISHB 225 at 46·8 kg/m

Area $=59\cdot66$ cm^2

$$r_{vv}=\frac{600}{4\cdot84}=124$$

Allowable stress $=670$ kg/cm^2

Increased allowable stress $=891$ kg/cm^2

Safe load on the column $=891\times 59\cdot66=53\cdot2$ t $>49\cdot5$ t

when the tank is empty maximum tensile force in the column $=11$ t.

2·10. Base Plate

Column load $=31$ t $=31,000$ kg

Area of base plate required $=\frac{31000}{40}=775$ cm^2

Use 37·5 cm × 25 cm base plate

Actual pressure on the base plate $=\frac{31000}{37\cdot5\times 25}=33$ kg/cm^2

Thickness of the base plate, $t=\sqrt{\frac{3w}{f_b}\left(A^2-\frac{B^2}{4}\right)}$

$$A=\frac{37\cdot5-22\cdot5}{2}=7\cdot5 \text{ cm}$$

$$B=\frac{25-22{\cdot}5}{2}=1{\cdot}25 \text{ cm}$$

$$\therefore \quad t=\sqrt{\frac{3\times 33}{1890}\left((7{\cdot}5^2-\frac{1{\cdot}25^2}{4}\right)}=\sqrt{{\cdot}05215\,(56{\cdot}61)}$$

$$=\sqrt{2{\cdot}92}=1{\cdot}65 \text{ cm}$$

Use 2 cm thick base plate.

2·10·1. Column Base Plate Connection

Use 70 mm×70 mm×6 mm cleat angles with two 20 mm ϕ rivets.

2·11. Force Horizontal Bacings

Total wind force on the tank and tower=3600+2940=6540 kg

Load on each column, $p=\frac{6540}{4}=1635$ kg

Horizontal shear in horizontal bracing$=\sqrt{2}\times 1635$

$$+\frac{\text{Column load}}{40}$$

$$=\sqrt{2}\times 1635+\frac{1}{40}\times 49500=2310+1240=3550 \text{ kg}$$

2·12. Diagonal Bracing

Length of diagonal$=\sqrt{3{\cdot}75^2+6^2}=\sqrt{14{\cdot}1+36}=7{\cdot}15$ m

Force in the diagonal$=3550\times\frac{7{\cdot}15}{3{\cdot}75}=6770$ kg

2·12·1. Design

Diagonal bracing
Load=6770 kg. (tension and comp)

Net area required$=\frac{6770}{1{\cdot}33\times 1500}=3{\cdot}38 \text{ cm}^2$

Try ISA 6565$-6{\cdot}5\times 6{\cdot}5\times 0{\cdot}6$ cm angle with 20 mm ϕ
Diameter of the rivet hole$=2{\cdot}0+0{\cdot}15=2{\cdot}15$ cm
Net area of the angle$=a+bk$

$$a=(6{\cdot}5-0{\cdot}3)\times 0{\cdot}6-2{\cdot}15\times 0{\cdot}6=3{\cdot}72-1{\cdot}29=2{\cdot}43 \text{ cm}^2$$

$$b=(6{\cdot}5-0{\cdot}3)\times 0{\cdot}6=3{\cdot}72 \text{ cm}^2$$

$$K=\frac{1}{1+0{\cdot}35\times\frac{3{\cdot}72}{2{\cdot}43}}=\frac{1}{1{\cdot}535}=0{\cdot}655$$

Net area provided$=a+bk=2{\cdot}43+0{\cdot}655\times 3{\cdot}72=2{\cdot}43+2{\cdot}42$
$=4{\cdot}85 \text{ cm}^2>3{\cdot}38 \text{ cm}^2$ required.

Hence use 65 mm×65 mm×6 mm angle

Horizontal bracing

Force in the bracing=3550 kg (comp)
Effective length$=0{\cdot}85\times 375=0{\cdot}85\times 375=318$ cm
Try $7{\cdot}5$ cm$\times 7{\cdot}5$ cm$\times 0{\cdot}6$ cm
$r_{min}=1{\cdot}46$ cm, Area$=8{\cdot}66 \text{ cm}^2$

$\frac{l}{r}=\frac{318}{1{\cdot}46}=217$

Allowable stress=220 kg/cm²
Increased stress=1·33×220=292 kg/cm²
Safe load=292×8·66=2530 kg<3550 kg. Hence unsafe.
Try 9·0 cm×9·0 cm×0·6 cm
Area of the section=10·47 cm²

$r_{min}=1{\cdot}75$
$l\ =318$ cm

$\frac{l}{r}=\frac{318}{1{\cdot}75}=182$

Allowable stress =330 kg/cm²
Increased stress =1·33×330=440 kg/cm²
Safe load =440×10·47=4600 kg>3550 kg Hence safe

2·13. Foundations Concrete Block

Total load on the column when the tank is full due to dead and wind loads=49·5 t

Self *wt* of foundation block (at 10%)=4·95 say 5 t
Total load on the column=54·5 t

Area of concrete base required$=\frac{54{\cdot}5}{15}=3{\cdot}63$ m²

Try 2 m×2 m×1 m. Size concrete block. Weight of concrete block=4×2200=8800 kg

Uplift force in the windward column when the tank is empty =11000 kg

Dead weight of concrete block should be atleast 1·5 times this load.

∴ dead weight of concrete block required=11000×1·5
=16,500 kg

Try 2·5 m×2·5 m×1·5 m
Dead weight of block=2·5×2.5×1·5×2200=20,600 kg
∴ Use 2·5 m×2·5 m×1·5 m block;

2·14. Anchor Bolts

Total uplift force=11000 kg
Using 4 No. of bolts

Force per bolt $=\frac{11000}{4}=2750$ kg

Net area of the bolt $=\frac{2750}{945\times1{\cdot}25}=2{\cdot}33$ cm²

Dia of bolt, $d=\sqrt{\frac{2{\cdot}33\times4}{\pi}}=1{\cdot}93$ cm

Use 4 Nos.—2 cm ϕ bolts.

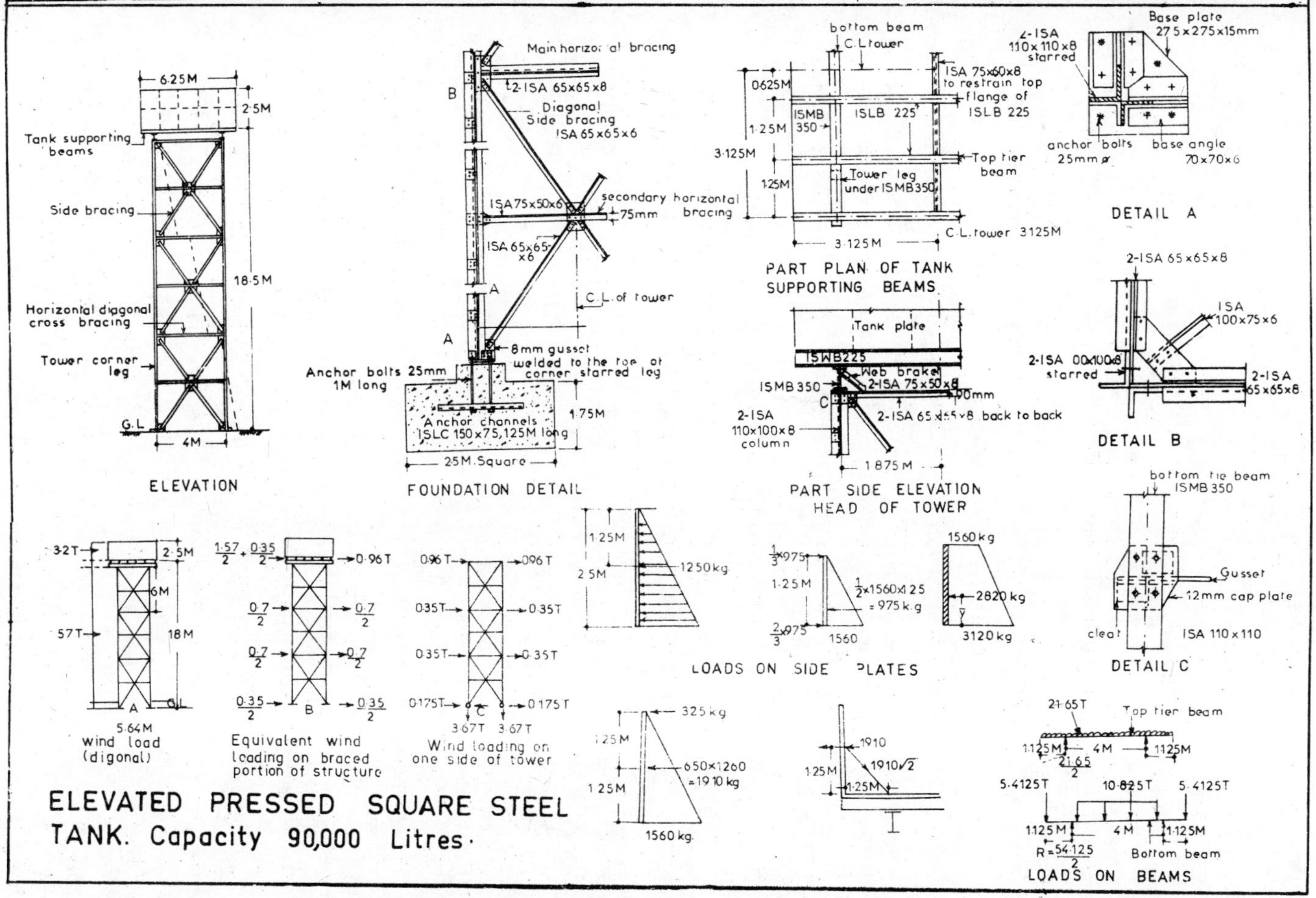

ELEVATED PRESSED SQUARE STEEL TANK. Capacity 90,000 Litres.

3. Pressed Steel Tank

3·1. Data

Capacity = 90 cu m; Free board = 15 cm

Height of the tank from GL to the bottoom of the tank = 18 m

Materials available; Pressed plates of size = 1·25 m × 1·25 m; Rolled sections

Bearing capacity of soil = 40 t/m²; Bearing on concrete = 40 kg/cm²; Wind load = 150 kg/m².

3·2. Relevant Codes

IS–804–1958; IS–875–1964; IS–800–1962; IS–496–1964

3·3. Trial Dimensions

Capacity of tank = 90 cu m; Depth of tank = 2·5 m

Depth of water = 2·35 m; Area in plan of the tank

$$= \frac{90}{2{\cdot}35} = 38{\cdot}25 \text{ m}^2$$

Side of the tank = 6·25 m; Use 5 plates of 1·25 × 1·25 m

Actual capacity provided = 6·25 × 6·25 × 2·35 = 92 cu m
> 90 cu m

3·4. Side Plates

Use 5 mm thick plates for top layer plates and 6 mm for bottom layer.

3·4·1. Water Loads

Water pressure at 1·25 m level = wh_1 = 1·25 × 1000 = 1250 kg/m
Water pressure at 2·5 m level = wh_2 = 2·5 × 1000 = 2500 kg/m
Pressure per plate width of 1·25 m = 1·25 × 1250 = 1560 kg
Total pressure on each plate = $\frac{1}{2}$ × 1560 × 1·25 = 975 kg

3·4·2. Loads on Top Plate

Force or reaction at the top of the plate $\frac{1}{3}$ × 975 = 325 kg
Force or reaction at bottom of the top plate $\frac{2}{3}$ × 975 = 650 kg

3·5. Loads on Bottom Plate

Water pressure per plate width at top = 1250 × 1·25 = 1560 kg

Water pressure per plate width at bottom of the bottom plate = 2500 × 1·25 = 3120 kg

Total load on the plate = $\frac{1}{2}$(1560 + 3120) × 1·25 = 2820 kg

Point of application of the pressure $\bar{y} = \frac{2 \times 3120}{1560 + 3120} \times \frac{1{\cdot}25}{3}$

$= 0{\cdot}558$ m

Force on the top of bottom plate due to water pressure in the bottom plate $=\frac{2820\times 0{\cdot}558}{1{\cdot}25}=1260$ kg

Force at bottom of the plate $=2820-1260=1560$ kg

Force at the mid depth at junction of top and bottom plate $=650+1260=1910$ kg

3·5·1. Stay

Force in the stay $=\sqrt{2}\times 1910=1{\cdot}41\times 1910=2700$ kg

Downward thrust at each vertical joint $=1910$ kg

This load should be considered in the design of tank beams.

3·6. Top Tier Beams

Capacity of tank $=6{\cdot}25\times 6{\cdot}25\times 2{\cdot}5=98$ cu m
Weight of water $=98$ t
Self weight of tank $=10$ t
Total load $=108$ t
Use 6 beams in the top tier

Assuming that end beams take only half the load carried by the intermediate beams

Load on the intermediate beam $=\frac{108}{5}=21{\cdot}6$ t

Self weight of beam (assume) $=50$ kg/m
Total load on the beam $=21{\cdot}65$ t
Centre to centre of column $=4$ m
That is the distance between the bottom tier girders $=4$ m
Force from the stay produces only cantilever moment

Maximum bending moment at centre $=\frac{21{\cdot}65}{2}\times 2-\frac{21{\cdot}65}{2}\times\frac{6{\cdot}25}{4}=21{\cdot}65-16{\cdot}9=4{\cdot}75$ t

Contilever moment due to load on the beam

$$=\frac{21{\cdot}65}{6{\cdot}25}\times\frac{1{\cdot}25}{2}=1{\cdot}95 \text{ mt}$$

Moment due to force from stay $=1{\cdot}95\times 1{\cdot}125=2{\cdot}15$ mt
Maximum cantilever moment $=1{\cdot}95+2{\cdot}15=4{\cdot}10$ mt
Maximum bending moment in the girder $=475000$ cm kg

Z = modulus of section required $=\frac{M}{f}=\frac{475000}{1575}=302\ \text{cm}^3$

Use ISWB 225×110 at 31·2 kg/m
Z provided $=305{\cdot}9\ \text{cm}^3$.

3·6·1. Shear Stress

Maximum shear force $=\frac{21{\cdot}65}{2}-\frac{21{\cdot}65}{6{\cdot}25}\times 1{\cdot}125=10{\cdot}825-3{\cdot}9$

$=6{\cdot}925$ t

$$\text{Shear stress} = \frac{6925}{\cdot 65 \times 17\cdot 33} = 619 \text{ kg/cm}^2 < 945 \text{ kg/cm}^2$$

3·7. Lower or Bottom Tier Beams

Reaction from top tier intermediate beam on to the bottom tier beam $= \frac{21\cdot 65}{2} = 10\cdot 825$ t

Reaction from edge beam $= \frac{10\cdot 825}{2} = 5\cdot 4125$ t

Total load on the girder $= 4 \times 10\cdot 825 + 2 \times 5\cdot 4125 = 54\cdot 125$ t

Bending moment at mid span

$$= \frac{54\cdot 125}{2} \times 2 - 5\cdot 4125 \times 3\cdot 125 - 10\cdot 825 \times 1\cdot 875 - 10\cdot 825$$

$$\times 0\cdot 675 = 54\cdot 125 - 16\cdot 9 - 19\cdot 65 - 7\cdot 35 = 11\cdot 1 \text{ mt}$$

Self weight of the beam (assume) $= 50$ kg/m

Maximum moment at centre of the beam

$$= \frac{\cdot 5 \times 6\cdot 25}{2} \times 2 - \frac{\cdot 05 \times 6\cdot 25}{2} \times \frac{6\cdot 25}{4}$$

$$= 0\cdot 3125 - 0\cdot 244 = 0\cdot 0685 \text{ mt} = 6850 \text{ cm kg}$$

Total maximum bending moment

$$= 111{,}000 + 6850 = 1116850 \text{ cm kg}$$

$$Z \text{ required} = \frac{1116{,}850}{1575} = 710 \text{ cm}^3$$

Use ISMB 350 × 165 at 49·5 kg/m

Z provided $= 751\cdot 9$ cm³

3·7·1. Shear Stress

Maximum shear force

$$= \frac{54\cdot 125 + \cdot 0495 \times 6\cdot 25}{2} - (\cdot 0495 \times 1\cdot 125 + 5\cdot 4125)$$

$$= 27\cdot 22 - 5\cdot 47 = 21\cdot 75 \text{ t}$$

$$\text{Shear stress} = \frac{21750}{28\cdot 83 \times 74} = 1020 \text{ kg/cm}^2 > 945 \text{ kg/cm}^2$$

Section is unsafe

Try ISWB 350 × 200, $t_w = 8$ mm, $h_i = 29\cdot 55$ cm, $Z = 887\cdot 9$ cm³ at 56·1 kg/m

$$\text{Shear stress} = \frac{21750}{\cdot 8 \times 29\cdot 55} = 922 \text{ kg/cm}^2 < 945 \text{ kg/cm}^2$$

3·8. Windloads Case—I

Elevation of the tank $= 18$ m above G.L.

Wind pressure $= 150$ kg/m²

3·8·1. Wind Load in Tank & Bearers

Wind load in the tank including supporting beams

$$= 6\cdot 25\,(2\cdot 5 + 0\cdot 225 + 0\cdot 350) \times 150 = 6\cdot 25 \times 3\cdot 075 \times 150$$

$$= 2890 \text{ kg} = 2\cdot 89 \text{ t}$$

This load is assumed to be distributed equally to windward and leeward frames.

Wind load on each side $= \frac{2 \cdot 89}{2} = 1 \cdot 445$ t

and acts at a height $= 17 \cdot 425 + \frac{3 \cdot 075}{2} = 18 \cdot 962$ m

3·8·2. Wind Load on the Frame

It is assumed that approximately 20% of one frame of the staging, is exposed to wind

Wind load on one frame $= 17 \cdot 425 \times 4 \times 150 \times \frac{20}{100} = 2 \cdot 1$ t

Assuming that the main horizontal bracings are provided at intervals of 6 m.

Equivalent load on the top of the staging, due to wind load on the tank $= 1 \cdot 445 \times \frac{18 \cdot 962}{17 \cdot 425} = 1 \cdot 57$ t

Wind load acting on the staging is distributed and assumed to act at the level of horizontal bracings. The intermediate bracings carry twice the load carried by the bracings at bottom and top

Total wind load on each frame $= 2 \cdot 1$ t

Total load at the level of intermediate bracings $= \frac{2 \cdot 1}{3} = 0 \cdot 7$ t

Wind load at the top and bottom bracing $= 0 \cdot 35$ t

In addition to the wind load of 0·35 t on the top bracing, a load of 1·57 t due to wind load on the tank, is also acting.

Therefore total wind load at top bracing $= 1 \cdot 57 + 0 \cdot 35 = 1 \cdot 92$ t

These loads are again distributed equally to act at all nodel points of the frame

Horizontal force at base level

$= 2(0 \cdot 96 + 0 \cdot 35 + 0 \cdot 35 + 0 \cdot 175) = 3 \cdot 67$ t

Taking moments about B the vertical reaction at the base due to wind loads

$$VA \times A = 2 \times 0 \cdot 96 \times 17 \cdot 425 + 2 \times 0 \cdot 35 \times 11 \cdot 425 + 2 \times 0 \cdot 35 \times 5 \cdot 425$$
$$= 46 \cdot 42$$

$$VA = \frac{46 \cdot 42}{4} = 11 \cdot 605 \text{ t}$$

3·9. Force In The Diagonal

Maximum force in the diagonal will occur in the bottom panel of the frame

Net shear force in the panel $= 3 \cdot 67 - 0 \cdot 35 = 3 \cdot 32$ t

Force in the diagonal member $= \frac{3 \cdot 32 \times \sqrt{5 \cdot 425^2 + 4^2}}{4}$

$$= 3 \cdot 32 \times \frac{6 \cdot 75}{4} = 5 \cdot 6 \text{ t}$$

3·10. Force In Main Horizontal Brace (Case—2)

Maximum force in the horizontal brace=3·32 t
Wind blowing diagonally
Wind load on the projected length of the diagonal of the tank
$=1·41\times6·25\times150\times0·8\ (2·5+0·225+0·35)$
$=1·41\times6·25\times120\times3·075=3200$ kg$=3·2$ t

A factor of 0·8 is considered, since some of the wind load on the tank is deflected off the sides of the tank.

Wind load on the tower across the diagonal on the projected area assuming 20% of exposed area

$$=1·41\times4\times17·425\times150\times\frac{20}{100}=2880 \text{ kg}=2·88 \text{ t}$$

Due to indeterminate and complex form of exposure on the open frame, reduction factor is not used

Wind load on wind ward frame=2·88 t

Wind load on leeward frame=2·88 t

Total wind load on the frame=5·76 t

3·11. Load in the Column due to Wind Loads

Overturning moment due to wind load

$$=3·2\times19·25+\frac{5·76\times17·425}{2}=61+50·2=111·2 \text{ mt}$$

$$\text{Force in the column}=\frac{111·2}{5·64}=19·7 \text{ t}$$

This force is compressive in leeward column and tensile in the wind ward column

3·12. Design of column

When tank is empty
Dead load of tank (assumed)=10 t
Self weight of staging =8 t
Load per column=$\frac{18}{4}$ =4·5 t (comp)
Maximum load in the widward column due to wind
=19·7 t (tension)
Net load on wind ward column
=19·7−4·5=15·2 t (tension)
Net load on leeward column=19·7+4·5=24·2 t (comp)
When the tank is full
Dead load per column=4·5 t

$$\text{Water load per column}=\frac{98}{4}=24·5 \text{ t}$$

Wind load=±19·7 t

Total load on wind ward column
=4·5+24·5−19·7=9·3 t (comp)
Total load on leeward column=4·5+24·5+19·7=48·7 t
Design load on the column
Tank empty=24·2 t (comp) and 15·2 t (tension)

Tank full+wind loads=48·7 t (comp)
Tank full and no wind load=29 t (comp)

3·12·1. Trial Section

110×110×8 mm, 2 equal angles starred

r_{vv}=4·28 cm, A=34·04 cm²

Subsidiary horizontal bracing is used at 3 m intervals in order to reduce the unsupported length of the columns.

Effective length of the column=3 m=300 cm

$$\frac{l}{r}=\frac{300}{4\cdot28}=70$$

Allowable stress=1096 kg/cm²

Stress induced$=\frac{29000}{34\cdot04}=850$ kg/cm². Hence safe

Stress induced under combined loads of wind and dead loads.

Maximum load on the column=48700 kg

Allowable stress=1096 kg/cm²

Increased stress when wind loads are considered
=1096×1·33=1480 kg/cm²

3·13. Base Plate Design

Column load=29 t=29000 kg

Area of base plate required$=\frac{29000}{40}=725$ cm²

Use 27·5 cm×27·5 cm

Actual pressure on the base plate$=\frac{29000}{27\cdot5\times27\cdot5}=38\cdot4$ kg/cm²

Thickness of the base plate

$$t=\sqrt{\frac{3w}{f_b}\left(A^2-\frac{B^2}{4}\right)}$$

$$A=\frac{27\cdot5-22\cdot0}{2}=\frac{5\cdot5}{2}\text{ cm}$$

$$B=\frac{27\cdot5-22\cdot0}{2}=\frac{5\cdot5}{2}\text{ cm}$$

$$t=\sqrt{\frac{3\times38\cdot4}{1890\times4}\left\{(5\cdot5)^2-\frac{(5\cdot5)^2}{4}\right\}}$$

$$=\sqrt{\frac{3\times38\cdot4\times22\cdot65}{1890\times4}}=\sqrt{1\cdot38}=1\cdot18\text{ cm}$$

Use 1·5 cm thick base plate.

For connecting column and the base plate.

Use 70 mm×70 mm×6 mm cleat angles with 2-20 mm ϕ rivets

Stress in the column$=\frac{48700}{34\cdot04}=1425$ kg/cm²<1480 kg/cm²

Hence safe

Use 110×110×8 mm 2 equal angles starred when the tank is empty, maximum wind load acts on the tank and the column is in tension.

The maximum tension = 15·2 t

The member however can easily take this tension. To prevent uplift, sufficient anchorage has to be provided with concrete block.

3·14. Diagonal Members

Maximum force in the diagonal = ±5·6 t

Length of the diagonal in the bottom panel = 6·75 m

Unsupported length of the diagonal $= \frac{6{\cdot}75}{2} = 3{\cdot}375$ m

$= 337{\cdot}5$ cm

Effective length = ·85 × 337·6 = 286 cm

$\frac{l}{r_{min}} \not> 250 \qquad \frac{l}{r_{min}} = \frac{270}{r_{min}} \not> 250$

r_{min} required $= \frac{286}{250} = 1{\cdot}15$ cm

Use ISA 65×65×6 mm which provides $r_{min} = 1{\cdot}26$ cm

3·15. Main Horizontal Bracing

Force is maximum in the bracing in bottom panel = 3·32 t (comp)

This number is required to carry the above force and provide restraint to corner legs. Since the force is only due to wind

l/r can be increased from 200 to 250

However overall rigidity is to be maintained however l/r is restricted to 180.

Try two angles strainer 75×75×8 mm

$r_{vv} = 2{\cdot}88$ cm $\qquad A = 22{\cdot}76$ cm²

$\frac{l}{r} = \frac{400 \times {\cdot}85}{2{\cdot}88} = 118 < 180$

Allowable stress = 726 kg/cm²

Increased stress = 1·33 × 726 = 965 kg/cm²

Allowable load = 965 × 22·76 = 22,000 kg > 3320 kg

Member is too safe

Try 65×65×8 mm $\qquad r_{vv} = 1{\cdot}97$ cm

$\frac{l}{r} = \frac{400 \times {\cdot}85}{1{\cdot}97} = 171 < 180$

Use 65×65×8 mm starred angles.

3·16. Secondary Horizontal Members

These members theoretically will not carry any loads and are used for stiffening the diagonal and vertical members. Usually unsymmetrical members.

Try ISA 75×50×6 mm, $\quad A = 7{\cdot}16$ cm²

Effective length = 2×0·85 = 1·70 cm

$$\frac{l}{r_{min}}=\frac{170}{1\cdot07}=159<180$$

Use ISA 75×50×6, keeping longer by outstanding from the face of the tower.

3·17. Horizontal Cross Bracing

The forces in these members are small and it is also difficult to assess them accurately. From the design point of view these may be taken as nominal bracing members which are introduced to offset any tendency for twist or torque to develop in the structure due to peculiar condition of loads.

Length of diagonal=1·41×4=5·64 m=564 cm

Member is partially restrained at mid point by the intersection with the opposing diagonal, unsupported length of diagonal

$$=\frac{5\cdot64}{2}=2\cdot82 \text{ m}$$

Effective length=1·25×2·82=3·52 m=352 cm

$l/r \ngtr 250$

$$r_{min} \text{ required}=\frac{352}{250}=1\cdot41 \text{ cm}$$

ISA 100×75×6 mm—provides 1·59 cm

3·18. Foundations

Total load on the column when the tank is full due to dead and wind load=48·7 t

Self weight of foundation block (10%)=5·3 t

Total load=48·7+5·3=54 t

$$\text{Area of concrete base required}=\frac{54}{40}=1\cdot35 \text{ m}^2$$

Uplift force in the windward column when tank is empty =15·27

Dead weight of concrete block should be at least 1·5 times this load.

Therefore dead weight of concrete block required=15·2×1·5 =22·8 t

Use 2·5 m×2·5 m×1·75 m

Dead weight=2·5 m×2·5 m×1·75×2·4=27·8 t>22·8 t

3·19. Anchor Bolts

Total uplift force=15200 kg

Using 4 number of bolts

$$\text{Force per bolt}=\frac{15200}{4}=3800 \text{ kg}$$

$$\text{Net area of the bolt}=\frac{3800}{945\times1\cdot25}=3\cdot22 \text{ cm}^2$$

$$\text{Diameter of bolt } d=\sqrt{\frac{3\cdot22\times4}{\pi}}=2\cdot02 \text{ cm}$$

Use 4—25 mm ϕ bolts.

4. Design of Elevated Rectangular Steel Water Tank

4·1. Data

Capacity = 200 cu m; Free board = 10 cm

Height of tank from GL to the bottom of lower tier beams = 15 m; Bearing capacity of soil = 20 t/m²; Bearing on concrete = 40 kg/cm²

Materials available; Plates of width = 1·25 m and above; Rolled and angle section of all sizes.

4·2. Relevant Codes

IS—804—1958; IS—875—1964; IS—800—1962.

4·2·1. Allowable Stresses

Plates in bending = 1650 kg/cm²; Rivets in shear = 945 kg/cm².

4·3. Trial Dimensions

Capacity = 200 cu m

Depth of water tank (using three plates of 1·35 m with 10 cm lap) = 3·75 m

Depth of water = 3·65 m

$$\text{Area of tank in plan} = \frac{200}{3\cdot65} = 54\cdot75 \text{ m}^2$$

Try 8·75 m × 6·25 m × 3·65 m size

Actual capacity provided = 200 cu m

Using 25 cm wide corner curved plates over all size of the tank = 9·25 m × 6·75 m × 3·65 m.

4·4. Tank Portion Bottom Plates

Consider 1 m width of plate spanning between two adjacent T-cover joints

Water load = 3·75 × 1000 = 3750 kg

Self weight of the plate (using 18 mm plates) = 158 kg

Total load per metre = w = 3750 + 158 = 3908 kg

$$\text{Bending moment} = \frac{Wl^2}{8} = 3908 \times (1\cdot25)^2 = 763\cdot28 \text{ m kg}$$

$$Z = \frac{76328}{1650} = 46\cdot32 \text{ cm}^3$$

Thickness of the plate

$$\tfrac{1}{6}\, bt^2 = 46\cdot32$$

$$t = 1\cdot67 \text{ cm}$$

Use 18 mm thick plates.

4·4·1. T-Cover Joint

The joint is continuous over the top tier beams.

Span $= 1{\cdot}25$ m

Load on the joint $= 1{\cdot}25 \times 1{\cdot}25 \times 3908 = 6100$ kg

Self weight of T-joint $= 65$ kg

Total load $= w = 6165$ kg

$$\text{Max. BM} = \frac{WL}{10} = \frac{6165 \times 1{\cdot}25}{10} = 770 \text{ m kg} = 77000 \text{ cm kg}$$

Try ISNT 150 and 15 cm × 1 cm plate as cover plate

Area of T-section $= 29{\cdot}08$ cm²

C.G. of T-section from base $= 3{\cdot}95$ cm

$$\text{C.G. of the arrangement} = \frac{15 \times 2{\cdot}8 \times 1{\cdot}4 + 29{\cdot}08(2{\cdot}8 + 3{\cdot}35)}{15 \times 2{\cdot}8 + 29{\cdot}08}$$

$$= \frac{254{\cdot}8}{71{\cdot}08} = 3{\cdot}6 \text{ cm}$$

$$\text{Moment of inertia} = I = \tfrac{1}{12} \times 15 \times (2{\cdot}8)^3 + 15 \times 2{\cdot}8 \left(3{\cdot}6 - \frac{2{\cdot}8}{2}\right)^2$$

$$+ 603{\cdot}8 + 29{\cdot}08\,(3\,95 - 0\,8)^2$$

$$= 27{\cdot}4 + 203{\cdot}5 + 603{\cdot}8 + 289 = 1123{\cdot}7 \text{ cm}^4$$

$$Z = \frac{I}{y} = \frac{1123{\cdot}7}{(17\,8 - 3{\cdot}6)} = \frac{1123{\cdot}7}{14{\cdot}2} = 79{\cdot}1 \text{ cm}^3$$

4·4·2. Stress in the Joint

$$\text{Stress in the joint} = \frac{M}{Z} = \frac{77000}{79{\cdot}1} = 975 \text{ kg/cm}^2 < 1575 \text{ kg/cm}^2$$

4·4·3. Rivet Design

$$\text{Shear force} = \frac{6165}{2} = 3082{\cdot}5 \text{ kg}$$

$$\text{Horizontal shear per cm} = \frac{SA\bar{y}}{I}$$

$$\text{Length of joint} = \frac{3082{\cdot}5\,(15 \times 2{\cdot}8 \times 2\,2)}{1123\,7} = 253{\cdot}5 \text{ kg}$$

Using 16 mm ϕ rivets

$$\text{Strength of one rivet in single shear} = \frac{\pi}{4} \times (1{\cdot}75)^2 \times 945$$

$$= 2273 \text{ kg}$$

Using 2—rivets in a row;

$$\text{pitch of rivets in a row} = \frac{2 \times 2273}{253{\cdot}5} = 18 \text{ cm}$$

Maximum pitch $= 10$ times the thickness of thinnest plate

$= 10 \times 1 = 10$ cm

Minimum pitch $= 2{\cdot}5\,d = 2{\cdot}5 \times 1{\cdot}75 = 4{\cdot}4$ cm

Use 16 mm ϕ rivets at 6 cm c/c.

4·5. Vertical Sides

Use 18 mm plate with joints similar to the T-cover joints with 10 mm thick cover plate. T-section will act as vertical stiffners. Assuming that stays are provided at mid depth of the tank in both the direetions connecting T-joints. BM on vertical stiffner at the level of stay:

$$=\frac{wh^3}{6}\times l=\frac{1000\times 1{\cdot}875^3\times 1{\cdot}25}{6}=1375 \text{ m kg}$$

$$=137500 \text{ cm kg}$$

$$\text{Stress in the joint}=\frac{M}{Z}=\frac{137500}{79{\cdot}1}=1730 \text{ kg/cm}^2>1575 \text{ kg/cm}^2$$

4·5·1. Stays

If the same T-joints are to be used, the stays must be used at two levels, cne at each horizontal joint of plates.

Water pressure on the side wall

$$P=\frac{1000\times 3{\cdot}75^2\times 1{\cdot}25}{2}=8770 \text{ kg}$$

$$\text{weight of side wall}=w_1=3{\cdot}75\times\frac{18}{100}\times 1{\cdot}25\times 7850=664 \text{ kg}$$

$$\text{weight of water contained in the corner}=W_2=0{\cdot}25\times 3{\cdot}75\times 1{\cdot}25\times 1000=1172 \text{ kg}$$

Let T be the tension in the top stay then 2 T is the tension in the bottom stay

Taking moments of all forces about 0,

$$T\times 2{\cdot}5+2T\times 1{\cdot}25=P\times 1{\cdot}25+W_1\,x_1+W_2\,x_2$$

$$5T=\tfrac{1}{2}\times 1000\times 3{\cdot}75\times 3{\cdot}75\times 1{\cdot}25\times 1{\cdot}25+664\times 0{\cdot}259+1172\times 0{\cdot}125=11299$$

$$T=\frac{11299}{5}=2260 \text{ kg}$$

Pull in the bottom stay=2260×2=4520 kg

$$\text{Area of cross section required for top stay}=\frac{2260}{1650}=1{\cdot}38 \text{ cm}^2$$

Use 20 mm ϕ for connection

Use 4 cm × 1 cm flat

For the bottom stay Use 6 cm × 1 cm.

4·6. Upper Tier Beam Loads

8 beams are used in the top tier parallel to thc shorter direction

Load for the T-cover joint on any intermediate beam=6165 kg

$$\text{Load for the end T-cover joint}=\frac{6165}{2}+W_1+W_2=3083+664+1172=4919 \text{ kg}$$

4·6·1. Bending Moments

Reaction due to concentrated loads from T-cover joints

$$=\frac{4{\cdot}919\times 2+6{\cdot}165\times 4}{2}=17{\cdot}249 \text{ t}$$

BM at support$=4{\cdot}919\times 1{\cdot}25=6{\cdot}15$ mt

$$\text{BM at centre}=17{\cdot}249\times\frac{3{\cdot}75}{2}-4{\cdot}919\times 3{\cdot}125$$

$$-6{\cdot}165\times 1{\cdot}875-6{\cdot}165\times 0\ 625=1{\cdot}54 \text{ mt}$$

BM due to self weight of the beam (assuming 35 kg/m)

$$=\frac{35\times 1{\cdot}5^2}{2}=39{\cdot}4 \text{ m kg}$$

$$\text{BM at centre}=35\times\frac{6{\cdot}75}{2}\times 1{\cdot}875-35\times\frac{6{\cdot}75}{2}\times\frac{6{\cdot}75}{4}$$

$$=22 \text{ m kg}$$

Total moment at centre$=1540+22=1562$ m kg

Total moment at support$=6150+39{\cdot}4=6189{\cdot}4$ m kg

$$Z \text{ required}=\frac{618940}{1650}=375 \text{ cm}^3$$

Use ISLB 275 at 33 kg/m

Z provided$=392{\cdot}4$ cm^3

$$\text{Maximum shear force}=17{\cdot}249+{\cdot}033\times\frac{6{\cdot}75}{2}-(4{\cdot}919+6{\cdot}165$$

$$+1{\cdot}5\times{\cdot}033)=6{\cdot}34 \text{ t}=6340 \text{ kg}$$

$$\text{Shear stress}=\frac{6340}{0{\cdot}64\times 27{\cdot}5}=360 \text{ kg/m}^2<945 \text{ kg/cm}^2$$

4·7. Bottom Supporting Beam Loads

Two beams are used in the direction perpendicular to the top tier beams. Each beam is continuous over 3 columns.

Reaction from the intermediate beam of the upper tier beam

$=17{\cdot}249+{\cdot}033\times 3{\cdot}375=17.36$ t

Reaction from end top tier beam$=4{\cdot}919\times 3+{\cdot}033\times 3{\cdot}375$

$=14{\cdot}868$ t

assume self wt of beam$=70$ kg/m

4·7·1. Bending Moments

Applying theorem of three moments.

$$M_A\, l_1\times 2\, M_B\,(l_1+l_2)+M_C\, l_2=-6\left[\frac{A_1\,x_1}{l_1}+\frac{A_2\,x_2}{l_2}\right]$$

$$M_A=M_c=-14{\cdot}868\times 1{\cdot}25-\frac{0\ 07\ (1{\cdot}5)^2}{2}=-18{\cdot}66 \text{ mt}$$

$$-18{\cdot}66\times 3{\cdot}125+2\, M_B\ (3{\cdot}125+3{\cdot}125)-18{\cdot}66\times 3{\cdot}125$$

$$=-\frac{6}{3{\cdot}125}\Big[\tfrac{1}{2}\times 1{\cdot}25\times 17{\cdot}44\times\frac{2\times 1{\cdot}25}{3}$$

$$+13{\cdot}19\times1{\cdot}25\left(1{\cdot}25+\frac{1{\cdot}25}{2}\right)+\frac{1{\cdot}25}{2}\times4{\cdot}25\times\left(1{\cdot}25+\frac{1{\cdot}25}{3}\right)$$

$$+\tfrac{1}{2}\times0{\cdot}625\times13{\cdot}19\left(2{\cdot}5+\frac{0{\cdot}625}{3}\right)+\tfrac{2}{3}\times3{\cdot}125\times{\cdot}0855\times\frac{3{\cdot}125}{2}\Bigg]$$

$$12{\cdot}5\ M_B=-97{\cdot}452$$

$$M_B=-7{\cdot}8 \text{ mt}$$

Maximum bending moment = 1866000 cm kg

$$Z \text{ required}=\frac{1866000}{1650}=1170 \text{ cm}^3$$

Use ISWB 400 at 66·7 kg/m

Z provided—1171·3 cm^3

4·7·2. Shear Force

Taking moment about B

$$R_A\times3{\cdot}125+7{\cdot}8-14{\cdot}868\times4{\cdot}375-17{\cdot}36\times3{\cdot}125$$

$$-17{\cdot}36\times1{\cdot}875-17{\cdot}36\times0{\cdot}625-0{\cdot}07\times4{\cdot}625\times\frac{4{\cdot}625}{2}=0$$

$$R_A=49{\cdot}7 \text{ t}$$

$$R_c=49{\cdot}7 \text{ t}$$

$$R_B=2\times14{\cdot}568+6\times17{\cdot}36+9{\cdot}25\times0{\cdot}07-2\times49{\cdot}7=35{\cdot}14 \text{ t}$$

Shear force at A or C

$$=49{\cdot}7-14{\cdot}868-17{\cdot}36-{\cdot}07\times1{\cdot}5=17{\cdot}367 \text{ t}=17367 \text{ kg}$$

Shear force at B

$$=35{\cdot}14-14{\cdot}868-17{\cdot}36+49{\cdot}7-17{\cdot}36-17{\cdot}36$$

$$-{\cdot}07\times4{\cdot}625=17{\cdot}57 \text{ t}=17570 \text{ kg}$$

$$\text{shear stress}=\frac{17570}{0{\cdot}86\times34{\cdot}05}=600 \text{ kg/cm}^2<945 \text{ kg/cm}^2$$

4·8. Staging

Elevation of the tank 15 m above GL

wind pressure = 150 kg/m^2

4·8·1. Wind Normal to Longer Side

Wind load on the tank and beam = 9·25 (3·75+0·275+0·4) ×150 = 6·15 t

Wind load on central frame (assuming 20% of area)

$$=6\times14{\cdot}325\times150\times\frac{20}{100}=2{\cdot}58 \text{ t}$$

Assuming that the main horizontal bracings are provided at intervals of 5 m, equivalent load on the top of the central frame of the staging due to wind load on the tank.

$$F\times14.325=\frac{6{\cdot}15}{2}\times16{\cdot}5375$$

$$F=\frac{3{\cdot}08\times16{\cdot}5375}{14{\cdot}325}=3{\cdot}56 \text{ t}$$

Total wind load on central frame = 2·58 t

Distributing this load to act at the nodel points of the frame

Load at intermediate node$=\frac{2\cdot58}{3}=\cdot86$ t

Load at top and bottom nodes$=\frac{\cdot86}{2}=\cdot43$ t

Total wind load at the top of central frame$=0\cdot43+3\cdot56$
$=3\cdot99$ t

4·8·2. Axial Load in the Column due to Wind Loads

Taking moments about B

$V_A\times3\cdot75=3\cdot99\times14\cdot325+0\cdot86\times9\cdot325+0\cdot86\times4\cdot325$
$=57\cdot3\times8\cdot02+3\cdot72=69\cdot04$

$$V_A=\frac{69\cdot04}{3\cdot75}=18\cdot4 \text{ t}$$

4·9. Column Design

Maximum load on the corner column due to water and self weight of tank and beams$=49\cdot7$ t

Axial load due to wind on the leeward column$=18\cdot4$ t

Total load$=49\cdot7+18\cdot4=68\cdot1$ t

When the tank is empty, load on the leeward column

$$=68\cdot1-\text{water load}=68\cdot1-\frac{225}{6}=68\cdot1-37\cdot5=30\cdot6 \text{ t}$$

Load on the windward column$=49\cdot7-37\cdot5-18\cdot4=-6\cdot2$ t

Self weight of column with bracing$=1\cdot9$ t (Say)

Maximum load on the column when the tank is full
$=68\cdot1+1\cdot9=70$ t

Minimum load on the column when the tank is empty
$=-6\cdot2+1\cdot9=-4\cdot3$ t

Effective length of column$=2\cdot5$ m$=250$ cm
(providing secondary horizontal bracings at mid height)

Try ISHB 200, $r_{min}=4\cdot51$ cm

$$\frac{l}{r_{min}}=\frac{250}{4\cdot51}=55\cdot5$$

Allowable stress$=1148\cdot9$ kg/cm²

increased allowable stress$=1\cdot33\times1148\cdot9$ kg/cm²

safe load on the column$=47\cdot54\times1\cdot33\times1148\cdot9$
$=73000$ kg
$=73 \text{ t}>70$ t

4·10. Base Plate

Load on the column$=70$ t

Area of base plate$=\frac{70000}{40\times1\cdot33}=1320$ cm²

Use 40 cm$\times$40 cm base plate

projection $A=B$ $=\frac{40\quad20}{2}=10$

pressure $w \quad = \dfrac{70000}{40 \times 40} 43.7 \text{ kg/cm}^2$

Thickness of base plate $t = \sqrt{\dfrac{3w}{f_b}\left(A^2 - \dfrac{B^2}{4}\right)}$

$$= \sqrt{\frac{3 \times 43{\cdot}7}{1890 \times 1\ 33}\left(10^2 - \frac{10^2}{4}\right)} - 1$$

$= 1{\cdot}99$ cm

Use $45 \times 45 \times 2$ cm base plate.

4·10·1. Base Plate Connection

Maximum load to be carried by one cleat angle $= \dfrac{70000}{2}$

$= 35000$ kg

Bearing area required $= \dfrac{35000}{1{\cdot}33 \times 1890} = 13{\cdot}9 \text{ cm}^2$

Using 40 cm length of angle width required $= \dfrac{13{\cdot}9}{40} = {\cdot}348$ cm

which is negligible

Use ISA $60 \times 60 \times 6$ mm angle with 20 mm ϕ rivets for connection

Strength of rivet in bearing $= 2{\cdot}15 \times 0{\cdot}6 \times 2125 \times 1{\cdot}25 = 3425$ kg

Number of rivets required $= \dfrac{4300}{3425} = 1{\cdot}26$

Use 2 Nos. of 20 mm ϕ.

4·11. Fouudation Block

Uplift force on the column $= 4{\cdot}3$ t

Total load on the foundation = column load + 10% for self weight $= 70 + 7 = 77$ t

Area of concrete base $= \dfrac{77}{20} = 3{\cdot}85 \text{ m}^2$

Use 2 m $\times$ 2 m $\times$ 75 m

Weight of concrete block $= 2 \times 2 \times {\cdot}75 \times 2{\cdot}3 = 6{\cdot}9$ t

Factor of safety against uplift when the tank is empty

$$= \frac{6{\cdot}9}{4\ 3} = 1{\cdot}61 > 1{\cdot}5$$

4·12. Anchor Bolt

Uplift force $= 4300$ kg

Using 4 bolts, force per bolt $= \dfrac{4300}{4} = 1075$ kg

Net area of the bolt $= \dfrac{1075}{1{\cdot}25 \times 945} = {\cdot}91 \text{ cm}^2$

diameter of bolt $d = \sqrt{\dfrac{{\cdot}91 \times 4}{\pi}} = 1{\cdot}08$ cm

Use 4—12 mm bolts per column.

4·13. Diagonal Member

Maximum force in the horizontal brace at GL$=6{\cdot}14-0{\cdot}43$
$=5{\cdot}71$ t

Force in the diagonal in the bottom panel of the staging

$$=5{\cdot}71 \times \frac{5{\cdot}72}{3{\cdot}75} = 8{\cdot}7 \text{ t}$$

Length of diagonal member$=5{\cdot}72$ m

Effective length of diagonal member $= \dfrac{0{\cdot}85 \times 5{\cdot}72}{2} = 243$ cm

$l/r \not> 250$

r_{min} required $= \dfrac{243}{250} = 0{\cdot}974$ cm

Use ISA $55 \times 55 \times 6$

r_{min} provided$=1{\cdot}06$ cm

Cross sectional area$=6{\cdot}26 \text{ cm}^2$

Net area required $= \dfrac{8700}{1{\cdot}33 \times 1500} = 4{\cdot}35 \text{ cm}^2$

Net area of the angle using 20 mm ϕ rivets for connection

$a=(5{\cdot}5-2{\cdot}15)\ 0{\cdot}6=2{\cdot}01$

$b=(5{\cdot}5-0{\cdot}6)\ 0{\cdot}6=2{\cdot}94$

$$K = \frac{1}{1+0{\cdot}35 \times \dfrac{2{\cdot}94}{2{\cdot}01}} = 0{\cdot}66$$

Net area provided$=2{\cdot}01+2{\cdot}94 \times 0{\cdot}66=5{\cdot}61 \text{ cm}^2 > 4{\cdot}35 \text{ cm}^2$

4·14. Horizontal Brace

Force in the brace$=5{\cdot}71$ t

Effective length $=3{\cdot}75 \times 0{\cdot}85=3{\cdot}19 \text{ m}=319$ cm

$l/r \not> 180$

Try ISA $100 \times 100 \times 6$

$A=11{\cdot}67 \text{ cm}^2$

$r_{min}=1{\cdot}95$ cm

$l/r_{min} = \dfrac{319}{1{\cdot}95} = 164{\cdot}6 < 180$

Allowable stress$=402{\cdot}3 \text{ kg/cm}^2$

Increased allowable stress$=1{\cdot}33 \times 402{\cdot}3=535 \text{ kg/cm}^2$

safe load$=535 \times 11{\cdot}67=6240 \text{ kg}=6{\cdot}24 \text{ t} > 5{\cdot}71$ t.

5. Design of a Cylindrical Steel Tank with Hemispherical Bottom

5·1. Data

Capacity=350 m^3; Elevation of the tank=18 m;
Free board=15 cm
Bearing capacity of soil=15 t/m^2
Bearing on concrete=40 kg/cm^2

5·2. Relevant Codes

IS—804—1958; IS—875—1964.

5·3. Trial Dimensions

Capacity=350 m^3
Depth of tank in the cylindrical portion=3·75 m
Depth of water in the cylindrical portion=3·60 m
Diameter of the tank= D meters

Radius of cylindrical portion= $\frac{D}{2}$ metres

Volume of cylindrical portion= $\frac{\pi \times D^2}{4} \times 3{\cdot}6 = 2{\cdot}82\ D^2$

Volume of water in hemispherical portion= $\frac{1}{2} \times \frac{4}{3}\ \pi\ r^3$

$$= \frac{1}{2} \times \frac{4}{3} \times \pi \times \left(\frac{D}{2}\right)^3 = 0{\cdot}262\ D^3$$

Total volume of cylindrical and hemispherical portions

$= 2{\cdot}82\ D^2 + 0{\cdot}262\ D^3$

Try $D = 8{\cdot}5$ m

Capacity provided $= 2{\cdot}82 \times 8{\cdot}5 + 0{\cdot}262 \times 8{\cdot}5^3 = 363\ m^3 > 350\ m^3$

Therefore use 3·75 m depth of tank with 8·5 m diameter in the cylindrical portion and a hemispherical portion of the same diameter.

5·4. Roof

Use 6 mm plates in the conical domed roof with angles ISA 50×50×6 as rafters fastened to the tank sides and to the top ring of 25 cm diameter.

Rise $= \frac{1}{6} \times 8{\cdot}5 = 1{\cdot}42$ m say 1·5 m
Use single riveted lap joints with 16 mm ϕ rivets
Minimum pitch $= 2{\cdot}5 \times 1{\cdot}75 = 4{\cdot}4$ cm
Maximum pitch $= 32 \times 0{\cdot}6 = 19{\cdot}2$ cm
Use a pitch of 10 cm c/c

5·5. Cylindrical Portion

Maximum pressure at the junction of cylindrical and hemisperical portion per metre height

$$p = wh = 1000 \times 3{\cdot}75 = 3700 \text{ kg}$$

$$\text{hoop tension} = p\frac{D}{2} = 3750 \times \frac{8{\cdot}5}{2} = 15900 \text{ kg}$$

$$T = \frac{pD}{2}$$

Area of the plate resisting this tension $= t \times 100$ cm²

$$t \times 100 \times f_s \times \eta = 15900$$

where η = efficiency of the riveted joint

f_s = allowable stress in steel kg/cm²

and t = thickness of the plate in cm

Assuming double riveted lap joints $\eta = 70\%$

$$t = \frac{15900}{100 \times 1000 \times 0{\cdot}7} = {\cdot}228 \text{ cm}$$

Use a minimum thickness of 6 mm for cylindrical plates.

5·5·1. Vertical Joints

These joints are designed for the hoop tension

Use double riveted lap joints with 16 mm diameter rivets

Strength of two rivets in single shear

$$= 2 \times \frac{\pi d^2}{4} f_s = 2 \times \pi \times \frac{1{\cdot}75^2}{4} \times 945 = 4545 \text{ kg}$$

Strength of rivet in bearing $= 2fb\, dt = 2 \times 2125 \times 1{\cdot}75 \times 0{\cdot}6$

$= 4462$ kg

Least strength of rivet $= 4462$ kg

Maximum hoop tension per sq metre at the junction $= 15900$ kg

$$\text{pitch of the rivets} = \frac{\text{Least strength}}{\text{Maximum hoop tension}} \times 100$$

$$= \frac{4462}{15900} \times 100 = 28{\cdot}2 \text{ cm}$$

Minimum pitch $= 2{\cdot}5\, d = 2{\cdot}5 \times 1{\cdot}75 = 4{\cdot}4$ cm

Maximum pitch $= 10 \times t = 10 \times 0{\cdot}6 = 6$ cm

(to prevent leakage through the joints)

Use 16 mm ϕ rivets at 6 cm c/c and double riveted lap joint

5·5·2. Horizontal Joints

These joints are designed to resist the dead load of the plates above the joints. The weight being small, use single riveted lap joints with 16 mm ϕ rivets at 6 cm c/c

5·6. Hemispherical Bottom

Maximum water pressure in the hemispherical portion

$$= w\,(h + h_1) = 1000\,(3{\cdot}75 + 4{\cdot}25) = 8000 \text{ kg/m}^2$$

$$\text{tension in hemispherical portion} = \frac{pD}{4} = \frac{8000 \times 8{\cdot}75}{4}$$

$$= 17500 \text{ kg/m}$$

$$\text{Thickness of the plate} = t = \frac{pD}{4 \times fs \times \eta \times 100}$$

$$=\frac{17500}{1002\times0{\cdot}7\times100}=.25 \text{ cm}$$

Use (minimum thickness) 6mm for plates

5·6·1. Radial & Circumferential Joints

Maximum hoop tension at the bottom $=\frac{pD}{4}=17500$ kg/m

Try single riveted lap joints using 16 mm ϕ rivets

Strength of a rivet in single shear $=\frac{\pi}{4}\times1{\cdot}75^2\times945=2273$ kg

Strength of a rivet in bearing $=f_b\times d\times t=2125\times1{\cdot}75\times0{\cdot}6$
$=2231$ kg

Least strength $=2231$ kg

pitch of the rivets $=\frac{2231}{17500}\times100=12{\cdot}8$ cm

Maximum pitch $=10t=10\times0{\cdot}6=6$ cm

Use single riveted lap joints with 16 mm ϕ rivets at 6 cm c/c

5·6·2. Connection of Cylindrical and Hemispherical portion

This joint has to carry all the dead loads carried by the hemispherical bottom

weight of water $=363000$ kg

self weight of plates in the hemispherlcal portion $=2\times\pi\times r^2\times t\times$weight of steel per m³

$$=2\times\pi\times\left(\frac{8{\cdot}5}{2}\right)^2\times\frac{6}{10\times100}\times7850=5340 \text{ kg}$$

weight of laps of plates & rivets in the joints at 25% of plate

$$=5340\times\frac{25}{100}=1335 \text{ kg}$$

Total weight $=363000+5340+1335=369675$ kg$=369{\cdot}675$ t

Load per metre circumference of the joint$=\frac{369675}{\pi\times8{\cdot}5}=13800$ kg

Try Double riveted lap joints with 16 mm ϕ rivets pitch of the rivets $=\frac{4462}{13800}\times100=32{\cdot}5$ cm c/c

Maximum pitch $=10t=10\times0{\cdot}6=6$ cm

Use Double revited lap joint with 16 mm ϕ rivets at 6 cm c/c

5·7. Columns

weight of water+hemispherical bottom+laps$=369{\cdot}675$ t

weight of plates in the cylindrical portion$=\pi\times D\times t\times7850\times h$

$$=\pi\times8{\cdot}5\times\frac{6}{10\times100}\times3{\cdot}75\times7850=4730 \text{ kg}$$

weight of laps and rivets at 25% of plate weight

$$=\frac{25}{100}\times4730=1182{\cdot}5 \text{ kg}$$

weight of roof plates $=\frac{\pi D}{2}\times$ slant height

$$=\frac{\pi}{2}\times 8{\cdot}5\times\sqrt{\left(\frac{D}{2}\right)^2+r^2}\times 7850\times t$$

$$=\frac{\pi}{2}\times 8{\cdot}5\times\sqrt{\left(\frac{8{\cdot}5}{2}\right)^2+45^2}\times 7850\times\frac{6}{10\times 100}=2850 \text{ kg}$$

weight of supporting (8 Nos 4·75 m) angles at 4·5 kg/m
$=4{\cdot}75\times 8\times 4{\cdot}5=171$ kg
Total weight on the columns
$=369{\cdot}675+4{\cdot}73+1{\cdot}1825+2{\cdot}85+0{\cdot}171\simeq 378{\cdot}608$ t
Try 8 columns

Load on each column $=\frac{378{\cdot}608}{8}=47{\cdot}32$ t

5·8. Wind Loads on Tank

Intensity of wind pressure $=150$ kg/m^2
shape factor $=0{\cdot}7$
wind load on the tank and roof = projected area $\times 0{\cdot}7\times 150$

$$=\left[h\times D+\tfrac{1}{2}\times D\times r+\tfrac{1}{2}\times\frac{\pi D^2}{4}\right]0{\cdot}7\times 150$$

$$=\left[3{\cdot}75\times 8{\cdot}5+\tfrac{1}{2}\times 8{\cdot}5\times 1{\cdot}5+\tfrac{1}{2}\times\frac{\pi}{4}\times(8{\cdot}4)^2\right]0{\cdot}7\times 150$$

$=7000$ kg

Assuming this acts at mid height of tank portion, point of application above G·L $=18+0{\cdot}5=18{\cdot}5$ m

5·8·1. Wind Loads on Staging

Wind loads on (*i*) column (20 cm wide) $=5\times 0{\cdot}2\times 150\times 18$
$=2700$ kg

(*ii*) Diagonal bracings (10 cm wide)
$=[12\times 0{\cdot}1\times\sqrt{6^2+3{\cdot}26^2}\cos 22°$
$+12\times 0{\cdot}1\times\sqrt{6^2+3{\cdot}26^2}\cos 67°]\times 150=1620$ kg

(*iii*) Horizontal bracings (20 cm wide)
$=6\times 3\times 0{\cdot}2\times 150+6\times 1{\cdot}27\times 0{\cdot}2\times 150=770$ kg

Total wind load on staging $=2700+1620+770=5090$ kg

This is assumed to act at mid height of the staging $=\frac{18}{2}=9$ m

5·8·2. Axial Load on Column Due to Wind

Axial load on the extreme column due to wind $=\frac{M_r}{\Sigma r^2}$

M = wind load moment $=7000\times 18{\cdot}5+5090\times 9=130000+45810$
$=175{,}810$ m kg
$\Sigma r^2=2\times 4{\cdot}25^2+4\times 3^2=72$

Axial load on the column $=\frac{175810\times 4{\cdot}25}{72}=10400$ kg $=10{\cdot}4$ t

Total maximum axial load on column due to wind and dead loads$=47{\cdot}32+10{\cdot}4=57{\cdot}72$ t

self weight of column (assume)$=1{\cdot}28$ t

Total load$=59$ t

5·9. Column Design

Maximum Load on the Leeward column$=59$ t

Effective length column$=6$ m$=600$ cm

$$\text{Approximate area required}=1{\cdot}5\times\frac{59000}{1250}=71 \text{ cm}^2$$

Try ISHB 250

$r_{yy}=5{\cdot}49$ cm $\quad A=64{\cdot}96$ cm^2

$$\frac{l}{r}=\frac{600}{5{\cdot}49}=110$$

Allowable stress$=753$ kg/cm^2

Increased allowable stress when wind loads are accounted $=1{\cdot}33\times753=1000$ kg/cm^2

safe load on the column$=1000\times64{\cdot}96$ t $>$ 59 t

5·10. Diagonal Members

Total wind load on tank and staging$=7000+5090=12090$ kg

5·11. Horizontal Brace

$$\text{horizontal force in each column}=P_h=\frac{12090}{8}=1510 \text{ kg}$$

$$\text{Force in horizontal bracing}=\frac{P_h}{\cos 67^\circ}+\frac{1}{40}\text{column load}$$

$$=\frac{1510}{0{\cdot}39}+\frac{59000}{40}=3870+1480=5350 \text{ kg (comp)}$$

Effective length$=0{\cdot}85\times326=277$ cm.

Try ISA $90\times90\times6$ mm

$A=10{\cdot}47$ cm^2 $\quad r_{min}=1{\cdot}75$ cm

$$\frac{l}{r_{min}}=\frac{297}{1{\cdot}75}=160$$

Allowable stress$=423$ kg/cm^2

Increased allowable stress$=1{\cdot}33\times423=564$ kg/cm^2

Safe load$=564\times10{\cdot}47=5890$ kg$>$5350 kg

5·12. Diagonal Brace

$$\text{Force in diagonal brace}=\frac{5350\times6{\cdot}84}{3{\cdot}26}=11200 \text{ kg}$$

$$\text{Net area required}=\frac{11200}{1{\cdot}33\times1500}=5{\cdot}7 \text{ cm}^2$$

Try ISA $70\times70\times6$ mm with 20 mm ϕ rivets for connection

Net area provided$=a+bk$

$a=7\times0{\cdot}6-2{\cdot}15\times0{\cdot}6=2{\cdot}9$

$b=6{\cdot}4\times{\cdot}6=3{\cdot}84$

$$k=\frac{1}{1+0.35\times\frac{3.84}{2.9}}=0.683$$

Net area provided $=2.9+0.683\times3.84=5.62\ cm^2$
Use ISA $75\times75\times6$

5.13. Column and Tank Wall Connection

Load on the column=47320 kg
Using 20 mm ϕ rivets, bearing strength$=0.6\times2.15\times2125=2740$ kg

$$\text{No. of rivets required to resist the load}=\frac{47320}{2740}=18$$

Use two angle sections with 9 rivets in each

5.14. Foundation Block

Dead load on the column when tank is empty; self wt of bottom hemisphere+25% laps+self wt of cylindrical portion+25% laps+roof plates+roof angles$=5340+1335+4730+1182.5+2850+171=15508$ kg

$$\text{Load per column}=\frac{15608}{8}=1960\ \text{kg}$$

self wt. of column (51 kg/m)$=51\times18=908$ kg
Self wt. of braces etc=132 kg
Total load on column=3000 kg
upward load in the column due to wind=10,400 kg
Net uplift force in the column$=10400-3000=7400$ kg
Total load on the foundation due to all loads when tank is full =column load+10% of column load$=59+5.9=64.9$ t say 65 t

$$\text{Area of concrete base}=\frac{65}{15}=4.33\ m^2$$

Try $2.2\times2.2\times1$ m concrete block
weight of concrete$=2.2\times2.2\times1\times2.3=11.65$ t

$$\text{Factor of safety against uplift when tank is empty}=\frac{11.65}{7.4}=1.51>1.5$$

Use $2.2\times2.2\times1$ m foundation block.

5.15. Anchor Bolts

uplift force=7400 kg

$$\text{Using four bolts, force per bolt}=\frac{7400}{4}=1850\ \text{kg}$$

$$\text{Net area of bolt}=\frac{1850}{1.25\times945}=1.531\ cm^2$$

$$\text{Diameter of bolt}=d=\sqrt{\frac{1.531\times4}{\pi}}=\sqrt{1.96}=1.4\ \text{cm}$$

Use 4—20 mm ϕ bolts per column.

CYLINDRICAL STEEL TANK WITH HEMISPERICAL BOTTOM

Capacity 350,000 Litres

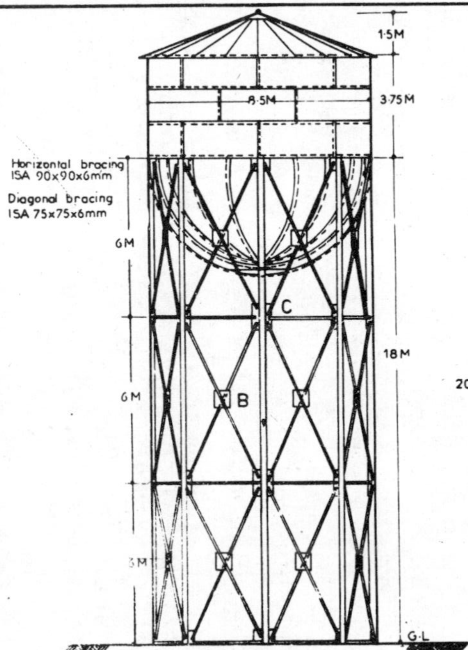

ELEVATION

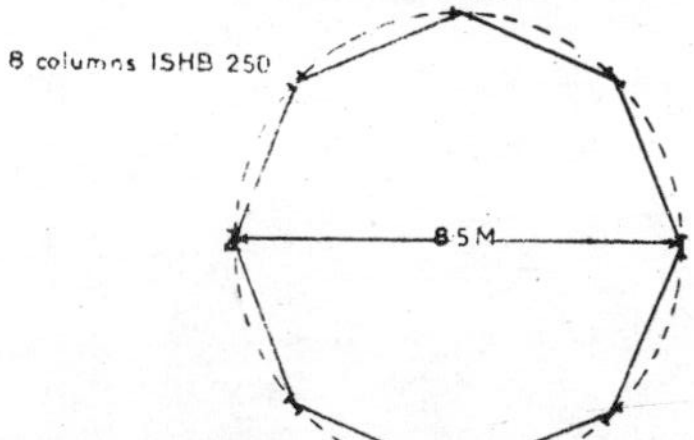

PLAN
SHOWING COLUMN ARRANGEMENT

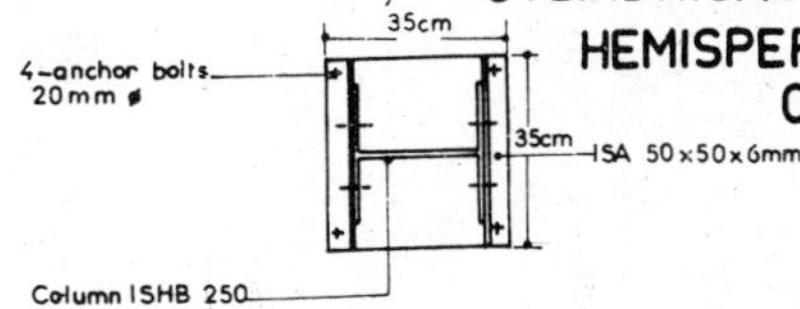

CONNECTION OF BASE PLATE WITH COLUMN

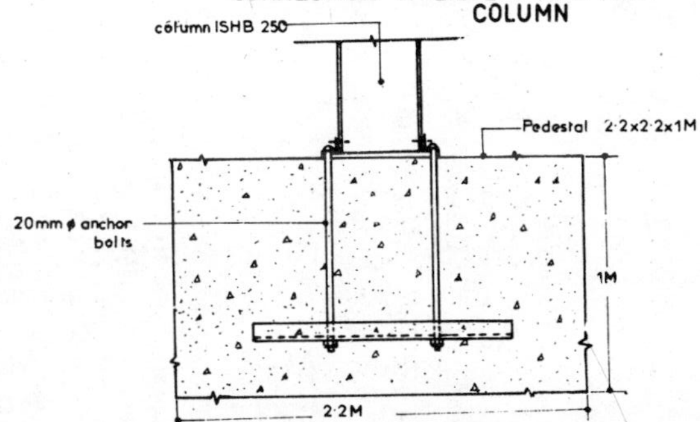

CONNECTION AT THE BASE WITH COLUMN

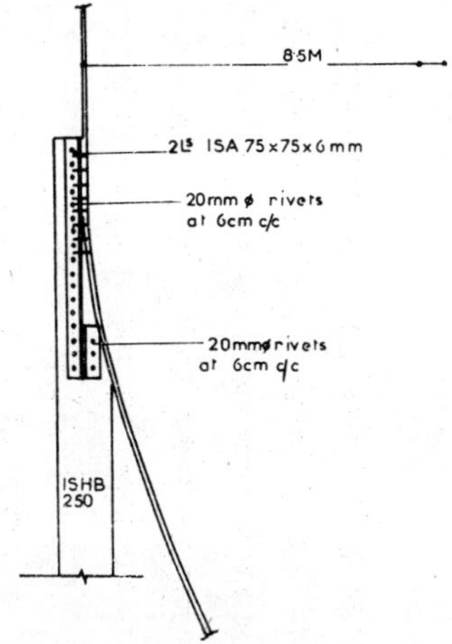

CONNECTION OF COLUMN WITH TANK WALL

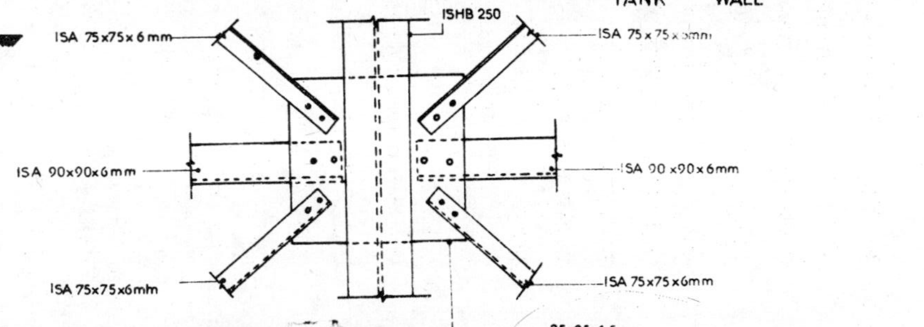

DETAILS OF JOINT C

5·16. Base Plate

Load on the base plate==59000 kg

Area of the base plate$=\frac{59000}{1·33\times 40}=1105 \text{ cm}^2$

Try 35×35 cm plate

pressure below the base plate$=\frac{59000}{35\times 35}=48·1 \text{ kg/cm}^2$

$$t=\sqrt{\frac{3w}{fb}\left(A^2-\frac{B^2}{4}\right)}=\sqrt{\frac{3\times 48·1}{1·33\times 1890}\left(25-\frac{25}{4}\right)}=1·04 \text{ cm}$$

Use $35\times 35\times 1·5$ cm thick base plate.

5·16·1. Connection of Colnmn to Base Piate

Use ISA $50\times 50\times 6$ mm cleat angles with 2—20 mm ϕ rivets.

6. Design of a Cylindrical Steel Tank with Conical Bottom

6·1. Data

Capacity $=200$ m³ ; Elevation of the tank $=20$m ;
Bearing capacity of Soil$=12$ t/m² ;
Bearing on concrete $=40$ kg/m²

6·2. Relevant Codes

IS—804—1958 ; *IS*—875—1964.

6·3. Trial Dimensions

Capacity $=200$ m³
D=Diameter of the cylindrical portion
h_1=the depth of cylindrical portion

$$\text{Volume of cylindrical portion} = \frac{\pi D^2}{4} h_1$$

D=Diameter of the conical portion
h_2=depth of conical portion

$$\text{Volume of conical portion} = \frac{1}{12} \pi D^2 h_2$$

Try $h_1=6$ m, $h_2=4$ m

$$\text{capacity} = \frac{\pi D^2 \times 6}{4} + \frac{1}{12} \pi D^2 \times 4 = 5{\cdot}726\ D^2 = 200$$

$D=5{\cdot}86$ m. Use $D=6$ m

$$\text{Actual volume provided} = \frac{\pi}{4} \times 6^2 \times 6 + 1/12 \times \pi \times 6^2 \times 4$$

$$= 206{\cdot}1 \text{ cu m} > 200 \text{ cu.m}$$

6·4. Roof

Provide conical domed roof with 6 mm plates. *ISA* 50×50×6 as rafters fastened to the side wall of the tank and to the top ring of 25 cm diameter.

Rise $=\frac{1}{4} \times 6 = 1{\cdot}5$ m

Use single riveted lap joints with 16 mm ϕ rivets at 10 cm c/c.

6·5. Cylindrical Portion

Plates are designed to resist hoop tension. Maximum water pressure at the junction of cylidrical wall and conical portion.

$P=wh=1000 \times 6=6000$ kg/m²

$$\text{Hoop tension} \quad T = p\frac{D}{2} = 6000 \times \frac{6}{2} = 18000 \text{ kg/m}$$

Area of the plate resisting this tension $= t\times 100$ cm^2, where
t = thickness of plate in cm

$t\times 100\times fs\times \eta = 18000$

where η = efficiency of the riveted joints
fs = allowable stress and steel

Using double riveted lap joints, $\eta = 0{\cdot}7$

$$t = \frac{18000}{100\times 1000\times 0{\cdot}7} = {\cdot}257 \text{ em}$$

Use 6 mm plates.

6·5·1. Vertical Joints

These joints are designed to resist the hoop tension. Try double riveted lap joints with 16 mm diameter rivets.

Strength of two rivets in single shear $= 2\times \frac{\pi d^2}{4}\times fs$

$$= 2\times\pi\times \frac{1{\cdot}75^2}{4}\times 945 = 4545 \text{ kg}$$

Strength of rivet in bearing $= 2 f_b dt = 2\times 2125\times 1{\cdot}75\times 0{\cdot}6$
$= 4462$ kg

Maximum hoop tension per sq. m $= 18000$ kg

Pitch of the rivets $= \frac{4462\times 100}{18000} = 24{\cdot}8$ cm

Minimum pitch $= 2{\cdot}5\times d = 2{\cdot}5\times 1{\cdot}75 = 4{\cdot}4$ cm

Maximum pitch $= 10\times t = 10\times {\cdot}6 = 6$ cm

Use double riveted lap joints with 16 mm ϕ rivets at 6 cm c/c

6·5·2. Horizontal Joints

These joints are designed to resist the dead load of the plates above the joints. The weight being small, use single riveted lap joints with 16 mm ϕ rivets at 6 cm c/c.

Thickness required to resist circumferential stress at the junction.

$$= \frac{p\, D \operatorname{cosec}\theta}{2\, t\times 1000\times 0{\cdot}7} = \frac{6000\times 6\times 1{\cdot}374}{2\times 100\times 1000\times 0{\cdot}7}$$

$= {\cdot}353$ cm, $<$ 6m provided

6·5·3. Circumferential Joints

These are designed to resist the hoop tension occuring in the radial direction.

Maximum hoop tension in the radial direction $= 12300$ kg/m

Try single riveted lap joints using 16 mm ϕ rivets.

Strength of a rivet in single shear $= \frac{\pi}{4}\times 1{\cdot}75^2\times 945 = 2273$ kg

Strength of a rivet in bearing $= f_b\, d\, t = 2125\times 1{\cdot}75\times 0{\cdot}6$
$= 2231$ kg

Pitch of the rivets $= \frac{2231}{12300}\times 100 = 18{\cdot}2$ cm

Maximum pitch $=10\ t=10\times0{\cdot}6=6$ cm

6·6. Conical Bottom

Maximum water pressure in the conical portion at the junction

$$=P=wh \quad \text{where } h=\left(h_1+h_2-\frac{D}{2}\tan\theta\right)$$

$$=\left(6+4-\frac{6}{2}\times\frac{4}{3}\right)=6 \text{ m}$$

$$p=1000\times6=6000 \text{ kg/m}^2$$

Max. hoop tension in the conical portion along the radial joints at the junction.

$$T=\frac{pD}{4}\ \text{cosec}\,\theta \text{ kg/m}$$

$$\text{Tan}\,\theta=\tfrac{4}{3}=1{\cdot}333$$

$$\sin\theta={\cdot}8007$$

$$\text{Cosec}\,\theta=1{\cdot}374$$

$$T=6000\times\tfrac{6}{4}\times1{\cdot}374=12300 \text{ kg/m}$$

Area of the plate resisting this tension along the radial joint $=t\times100$ cm^2

If the efficiency of the joint$=0{\cdot}7$

$$t\times100\times0{\cdot}7\times fs=T$$

$$t=\frac{12300}{100\times0{\cdot}7\times1000}=0{\cdot}18 \text{ cm}$$

Use 6 mm thick plates as provided for the circumferential hoop tension at the junction.

6·6·1. Radial Joints

These joints are designed to resist hoop tension in the circumferential direction.

Maximum hoop tension in the circumferential direction

$$=\frac{6000\times6}{2}\times\text{Cosec}\,\theta=18000\times1{\cdot}374=24700 \text{ kg/m}$$

Try single riveted hoop joints using 16 mm ϕ rivets

$$\text{Pitch of the rivets}=\frac{2231}{24700}\times100=9 \text{ cm c/c}$$

Use 16 mm ϕ at 6 cm c/c—single riveted.

6·7. Connection of Cylindrical and Conical Bottom

A ring is provided to take force due to the horizontal component of radial tension the conical portion.

Radial tension in the conical portion

$$T_c=P\frac{D}{4}\ \text{Cosec}\,\theta \text{ kg/m}=6000\times\tfrac{6}{4}\times1{\cdot}374=12300 \text{ kg/m}$$

Horizontal component of this force$=T_c\cos\theta=12300\times0{\cdot}6004$

$$=7400 \text{ kg/m}$$

Compressive force induced in the ring due to this horizontal component$=T_c \cos\theta \times \frac{D}{2}$ kg$=7400\times\frac{6}{2}=22200$ kg/m width of ring

6·8. Ring Design

The ring is designed as a strut, having slenderness ratio not less than $=\frac{\pi D}{2\sqrt{3r}}$

where r is the radius of gyration of the ring beam section about vertical axis.

Try *ISA* 100×100×6 mn
$A=11{\cdot}67$ cm^2, $r_y=3{\cdot}09$ cm

$$\text{Slenderness ratio of the strut}=\frac{\pi\times 600}{2\times 1{\cdot}73\sqrt{3{\cdot}09}}$$

$$=\frac{\pi\times 600}{2\times 1{\cdot}73\times 1{\cdot}75}=312$$

Allowable stress$=101$ kg/cm^2
Safe load/Meter of ring$=101\times 11{\cdot}67\times 100$
$=119000 \text{ kg} > 22{,}200 \text{ kg}$

Angle section less than this can be used since it is oversafe. However l/r ratio should not be greater than 350.

Use 16 mm ϕ, single riveted joints to connect the two portions using the angle ring.

$$\text{Pitch of the rivets}=\frac{2231}{24700}\times 100=9 \text{ cm c/c}$$

Use 16 mm ϕ at 6 cm c/c

6·9. Circular Girder Loads

Circular girder has to take entire self weight of tank and its contents. It is supported on a number of columns.

Weight of water $=206100$ kg$=206{\cdot}1$ t
Weight of roof $=\pi\, r\, s\times 7850\times t$

$$=\pi\times 3\times\sqrt{3^2+1{\cdot}5^2}\times\frac{0{\cdot}6}{100}\times 7850$$

$$=1480 \text{ kg}$$

Weight of sides $=\pi\times D\times t\times h_1\times 7850$

$$=\pi\times 6\times\frac{0{\cdot}6}{100}\times 6\times 7850=5320 \text{ kg}$$

$$\text{Weight of conical bottom}=\pi\times 3\times\sqrt{3^2+4^2}\times\frac{0{\cdot}6}{100}\times 7850$$

$$=2220 \text{ kg}$$

Total self weight of tank $=1480+5320+2220=9020$ kg
Rivets, roof angles, ring at the junction (at 25%)

$$=9020\times\frac{25}{100}=2260 \text{ kg}$$

Total self weight $=9020+2260=11280$ kg$=11{\cdot}28$ t
Load on the ring girder

$$W=206{\cdot}1+11{\cdot}28=217{\cdot}38 \text{ t}$$

Assuming self weight of girder including connections, at 150 kg/m

6·10. Ring Beam Desing

Weight of girder (including connections) $=\pi \times D \times 150$
$=\pi \times 6 \times 150=2820$ kg$=2{\cdot}82$ t
Total vertical load on the girder $=217{\cdot}38+2.82=220{\cdot}20$ t
Say$=221$ t

6·10. Bending Moment

Using 4 columns.

Maximum bending moment at support $=0{\cdot}0342\ W_r$
$=0{\cdot}0342 \times 221 \times \frac{6}{2}=22{\cdot}75$ mt
Maximum BM at centre$=0{\cdot}0176\ W_r=0{\cdot}0176 \times 221 \times \frac{6}{2}=11{\cdot}7$ mt
Torsional moment $=0{\cdot}0053\ W_r=0{\cdot}0053 \times 221 \times \frac{6}{2}=3{\cdot}52$ mt
Length of girder between two supports, that is unsupported length of girder $=\frac{\pi D}{4}=\frac{\pi}{4} \times 6=4{\cdot}72$ m

Modulus of section required on bending moment consideration

$$=\frac{M}{f}=\frac{2275000}{1653}=1380 \text{ cm}^3$$

Since the ring beam is subjected to torsional moment in addition to the bending moment.

Try section ISMB$=600$
Span $=4{\cdot}72$ m
Permistible stressess for a span of
4·5 m —1092 kg/cm²
5·00 m—995 kg/cm²
and for 4·72 m—1049 kg/cm²

Modulus of section required $=\frac{2275000}{1049}=2170 \text{ cm}^3$

Z_{xx} provided. $3060{\cdot}4 \text{ cm}^3$
Section is over safe.
Try ISMB-550, $Z_{xx}=2359{\cdot}8 \text{ cm}^3$
Span$=4{\cdot}72$ m

Permissible stress $=1005-(1005-914) \times \frac{\cdot 22}{\cdot 5}=965 \text{ kg/cm}^3$

Required modulus of section $=\frac{227500}{965}=2350 \text{ cm}^3$

Z_{xx} provided$=2359{\cdot}8 \text{ cm}^3$.

6·11. Columns

Total axial load per colum, due to dead and water loads

6·11·1. Winds Loads

Wind load on the projected area of roof and tank projected area $=\frac{1}{2}\times6\times1·5+6\times6+\frac{1}{2}\times4\times6=4·5+36+12=52·5$ sq m

Wind load $=0·7\times150\times52·5=5510$ kg $=5·51$ t

Assuming that this load acts at mid height of the cylindrical portion, the point of application. $=20+3=23$ m from G.L.

Wind load on staging at 20% of projected area $=0·2\times6\times20\times150=3600$ kg $=3·6$ t

Point of application of this force $=10$ m above G.L.

wind load moment $=5·51\times23+3·6\times10=127+36=163$ m t

Axial load per column due to wind load $=\dfrac{M_r}{\Sigma r^2}=\dfrac{163\times3}{2\times3^2}$

$$=\frac{163\times3}{2\times3\times3}=27·2 \text{ t}$$

Self weight of the column at 100 kg/m $=20\times100=2000$ kg $=2$ t

Maximum total axial load on the column $=55·25+27·20+2$ $=84·45$ t

Assuming horizontal braces at 5 m intervals unsupported length of column $=5$ m $=500$ cm

6·12. Column Design

Try ISHB—250

$A=64·96$ cm^2

$r_{min}=5·37$ cm

$$\frac{l}{r}=\frac{500}{5·37}=93$$

Allowable stress $=902·6$ kg/cm^2

Increased allowable stress $=1·33\times902·6$ kg/cm^2

Area required $=\dfrac{84450}{902·6\times1·33}=70·1$ sq cm.

Section is not sufficient

Try ISHB 300

$A=75·84$ cm^2

$r_{min}=5·41$ cm

$$\frac{l}{r}=\frac{500}{5·41}=92$$

Allowable stress $=910·4$ kg/cm^2

Increased allowable stress $=1·33\times910·4=1220$ kg/cm^2

Area required $=\dfrac{84450}{1220}=69·1$ sq cm $<74·85$ sq cm provided

6·13. Diagonal Bracing

Total wind force $=5·51+3·6=9·11$ t

Horizontal shear per column $=\dfrac{9·11}{4}=2·28$ t

Horizontal shear in the bracing $=2{\cdot}28\sqrt{2}+\dfrac{\text{column load}}{40}$

$=3{\cdot}3+\dfrac{84{\cdot}45}{40}=3{\cdot}3+2{\cdot}11=5{\cdot}41$ t

Force in the diagonal bracing

$=5{\cdot}41\times\sqrt{\dfrac{(3\sqrt{2})^2+5^2}{3\sqrt{2}}}=5{\cdot}41\times1{\cdot}34=7{\cdot}3$ t

Area required $=\dfrac{7300}{1{\cdot}33\times1500}=3{\cdot}64\text{ cm}^2$

Try ISA 65×65×6 mm with 20 mm ϕ rivets for connection.
Diameter of the rivet hole $=2+0{\cdot}15=2{\cdot}15$ cm
Net area of the angle $=a+bk$
$a=(6{\cdot}5-0{\cdot}3)\ 0\ 6-2{\cdot}15\times0{\cdot}6=2{\cdot}43\text{ cm}^2$
$b=(6{\cdot}5-0{\cdot}3)\ \times0{\cdot}6=3{\cdot}72\text{ cm}^2$

$k=\dfrac{1}{1+0{\cdot}35\times\dfrac{3{\cdot}72}{2{\cdot}43}}=0{\cdot}655$

Net area provided $=a+bk=2{\cdot}43+0{\cdot}655\times3{\cdot}72=2{\cdot}43+2{\cdot}42$
$=4{\cdot}85\text{ cm}^2>3{\cdot}64\text{ cm}^2$

6·14. Horizontal Bracing

Force in the horizontal bracing $=5{\cdot}41$ t (comp)
Length of the member $=3\sqrt{2}=4{\cdot}25$ m
Effective length $=0{\cdot}85\times4{\cdot}25=360$ cm
Try ISA 100×100×8 with 20 mm ϕ rivets for connection

$A=15{\cdot}39\text{ cm}^2$

$r_{min}=1{\cdot}95$ cm

$\dfrac{l}{r_{min}}=\dfrac{360}{1{\cdot}95}=185$

Increased allowable stress$=1{\cdot}33\times318=424\text{ kg/cm}^2$
Safe load$=424\times15{\cdot}39=6{\cdot}5\text{ t}>5{\cdot}41$ t

6·15. Foundation

Dead load on each column when the tank is empty

$=\dfrac{221-206{\cdot}1}{4}=3{\cdot}72$ t

Self weight of each column$=2{\cdot}0$ t
Upward axial force in each column due to wind$=27{\cdot}20$ t
Net upward force in each column$=27{\cdot}2-5{\cdot}72$ t $=21{\cdot}48$ t
Maximum axial load in the column when the tank is full $=84\ 85$ t

Area of concrete base required $=\dfrac{84{\cdot}85}{12}=7{\cdot}05$ sq m

Try $1{\cdot}75\times2{\cdot}75\times1{\cdot}5$ m concrete block

Bearing capacity$=\dfrac{84{\cdot}85}{2{\cdot}75\times2{\cdot}75}=8{\cdot}5\text{ t/m}^2<12\text{ tm}^2$

Weight of the concrete block $= 2{\cdot}75 \times 2{\cdot}75 \times 1{\cdot}5 \times 2300$
$= 26{\cdot}2 \text{ t} > 21{\cdot}48 \text{ t}$

Factor of safety against uplift force $= \dfrac{26{\cdot}2}{21{\cdot}48} = 1{\cdot}23 < 1{\cdot}5$

Increase the thickness of the block to 1·75 m
Dead weight of block $= 2{\cdot}75 \times 2{\cdot}75 \times 1{\cdot}75 \times 2300 = 30{\cdot}5 \text{ t}$

F.S against uplift $= \dfrac{30{\cdot}5}{21{\cdot}48} = 1{\cdot}57 > 1{\cdot}5$

6·16. Base Plate

Area of base plate required $= \dfrac{8485}{1{\cdot}33 \times 40} = 1590 \text{ cm}^2$

Provide 40×40 cm base plate. Pressure below the plate

$$= \frac{8485}{1600} = 53 \text{ kg/cm}^2$$

$f_b = 1{\cdot}33 \times 1890 = 2520 \text{ kg/cm}^2$

$$A = \frac{40-30}{2} = 5\text{cm}$$

$$B = \frac{40-30}{2} = 5 \text{ cm}$$

$$t = \sqrt{\frac{3 \times 53}{2520}\left(5^2 - \frac{5^2}{4}\right)} = 1{\cdot}90 \text{ cm}$$

Use 15 mm thick plate with Four anchor bolts.

Force per bolt $= \dfrac{21{\cdot}48}{4} = 5{\cdot}37 \text{ t}$

Area per bolt required $= \dfrac{5370}{1{\cdot}25 \times 945} = 4{\cdot}37 \text{ cm}^2$

Use 25 mm dia bolts.

6·17. Connection

Use 75×75×6 mm cleat angle to connect base plate to column with 20 mm ϕ rivets

Bearing strength of rivet $= 1{\cdot}25 \times 2{\cdot}15 \times 0{\cdot}6 \times 2125 = 3426 \text{ kg}$

No. of rivets required $= \dfrac{21480}{3426} = 6{\cdot}25$

Use 8 Nos with gusset plate

PART V

DESIGN OF STEEL CHIMNEYS

1. Guyed Steel Chimneys General Information

1·1. Size

Before designing a large chimney, height and diameter of the chimney must be decided. The height should be so selected to give the draft required and the diameter must be large enough to permit passage of gases.

1·2. Types

(*i*) Guyed steel stacks usually unlined (*ii*) Self supporting stacks usually lined.

1·3. Loads

(*i*) Weight of the chimney and lining, if any (*ii*) Foundation reactions (*iii*) Wind pressure (*iv*) The pull of gauge in the case of guyed stacks (*v*) Forces due to earth quake shocks.

1·4. Breech Opening

This opening is about 20% larger than the internal cross sectional area of the chimney. The maximum width of the breech opening should not exceed $\frac{2}{3}$ of the diameter D.

In guyed steel stacks the breech opening is often below the base of the stack and in the case of self-supported chimney the breech opening should be far enough above the conical base.

1·5. Guyed Steel Chimney Types

(*i*) Chimney with one set of guys consisting of 3 or 4 or sometimes 6 wires attached at a height of $\frac{1}{4}$ or $\frac{1}{3}$ from top.
(*ii*) Chimneys with two sets of guys consisting of 3 or 4 wires attached at different heights.
(*iii*) Chimney with three sets of guys.
(*iv*) Chimneys in a continuous row with lattice bracing between them.

1·5. Design

In the design of guyed chimneys the following loads must be considered.

1·6·1. Minimum Thickness of Plates

For diameter over 1 m, a minimum thickness of 6 mm is recommended.

1·6·2. Self Weight

The self weight of the chimney

$W_s = D \times H \times t = 7850$ kg

where

D=diameter of the chimney in m
H=height of the chimney in m
t=thickness of steel plate of the chimney wall in m

To this weight 15% extra is added to account for laps in the plates and rivets in vertical and horizontal seams.

1·6·3. Pull on Guy Wires

The guy wires must be designed to take care of entire wind reaction at a collar. The maximum pull in any guy wire will occur when the wind blows along that guy wire.

$$\text{Pull in guy wire} = P_g = \frac{R_c}{\sin\theta} = R_c \operatorname{cosec}\theta$$

The vertical component of the pull in the guy wire

$$\frac{R_g}{P_g} = \cos\theta$$

$$R_g = P_g \cos\theta = \frac{R_c}{\sin\theta}\cos\theta = R_c \cot\theta$$

θ varies from 30° to 75° to the vertical in most of the cases.

Initial stress in guy wires may be assumed to be=350 kg/cm²

1·7. Wind Moments

(*i*) Chimneys with one set of guys

In the analysis it is assumed that the chimney is considered as a beam, simply supported at the base and at the collar where the guys are attached.

1·7·1. Shear Forces

Reaction at the collar level

Taking moments about the base

$$R_c \times h_1 - P_w \times \frac{H}{2} = 0$$

$$R_c = \frac{P_w H}{2h_1}$$

$$\text{Shear at the base } R_b = P_w - R_c = P_w - \frac{P_w H}{2h_1}$$

$$\text{Shear just above the collar} = \frac{P_w}{H}(H-h_1)$$

$$\text{Shear just below the collar} = R_c - \frac{P_w}{H}(H-h_1)$$

1·7·2. Bending Moments

$$\text{BM at collar level} = -\frac{P_w}{H}\left(H-h_1\right)\left(\frac{H-h_1}{2}\right) = -\frac{P_w}{2H}\left(H-h_1\right)^2$$

Maximum BM between the base and collar will occur at point of zero shear

$$R_b - \frac{P_w}{H}\times l = 0$$

$$l=\frac{R_b H}{P_w}+\left(P_w-\frac{P_w H}{2h_1}\right)\frac{H}{P_w}=\left(H-\frac{H^2}{2h_1}\right)\text{above base}$$

Maximum BM at a height x from base$=R_b l=\frac{P_w}{H}\times l\times\frac{l}{2}$

$$=\left(P_w-\frac{P_w H}{2h_1}\right)\left(H-\frac{H^2}{2h_1}\right)-\frac{P_w}{2H}\times\left(H-\frac{H^2}{2h}\right)^2$$

$$=\frac{P_w}{H}\left(H-\frac{H^2}{2h_1}\right)\left(H-\frac{H^2}{2h_1}\right)-\frac{P_w}{2H}\left(H-\frac{H^2}{2h}\right)^2$$

$$=\left(\frac{P_w}{H}-\frac{P_w}{2H}\right)\left(H-\frac{H^2}{2h_1}\right)^2=\frac{P_w}{2H}\left(H-\frac{H^2}{2h_1}\right)^2$$

$$=\frac{P_w}{2}\left(1-\frac{H}{2h_1}\right)^2$$

1·8. Case (i) Guys At H/3

When the guys are at $\frac{1}{3}H$ from top of the chimney

$$(H-h_1)=\frac{H}{3}$$

Substituting

1·8·1. Shear Force

SF at base$=-\frac{P_w}{4}$

SF just above the collar$=-\frac{P_w}{3}$

SF just below the collar$=\frac{5}{12}P_w$

$R_c=\frac{3}{4}P_w$ and $R_b=\frac{P_w}{4}$

1·8·2. Bending Moments

The BM at the collar level$=\frac{P_w}{H}\times\frac{H}{3}\times\frac{H}{6}=\frac{P_w H}{18}$

BM at the point of zero shear$=\frac{P_w H}{2}\left(1-\frac{H\times 3}{2\times 2H}\right)^2=\frac{P_w H}{32}$

In the design, however, all the sections below the collar should be designed to carry a moment of $\frac{P_w H}{18}$

1·9. Case (ii) Guys at $\frac{H}{4}$

Guys at $\frac{H}{4}$ below top

$$(H-h_1)=\frac{H}{4}$$

$R_c=\frac{2}{3}P_w$ and $R_b=\frac{P_w}{3}$

1·9·1. Shear Forces

SF just above the collar $= -\dfrac{P_w}{4}$

SF just below the collar $= \dfrac{5}{12} P_w$

SF at base $= -\dfrac{P_w}{3}$

Point of zero shear $l = \dfrac{H}{3}$

1·9·2. Bending Moments

BM at the collar $= \dfrac{P_w H}{3^2}$

BM at the point of shear $= \dfrac{P_w H}{18}$

Design moment for the section between the collar and a point $\dfrac{H}{3}$ from the base $= \dfrac{P_w H}{32}$ at collar varying to $\dfrac{P_w H}{18}$ at $\dfrac{H}{3}$ from the base and for all sections below $\dfrac{H}{3}$ from the base $= \dfrac{P_w}{18}$

1·10. Chimneys with two sets of Guys

Approximate method—1. In the analysis, it is assumed that the upper collar carries the wind load from the top of the chimney to a point midway between the two collars and that the second collar carries the load from the point to a point midway between the lower collar and the base and that the base carries the remainder of the load.

Approximate method—2. It is assumed for determining the reactions, shears and moments due to the wind, that

(*i*) Upper collar carries the wind load from the top of the chimney to a point at a distance below the top of the chimney equal to 2 times the distance from the top of the stack to the collar plus half the remaining distance to the lower collar.

(*ii*) The lower collar carries the wind load from this point to a point midway between the lower collar and the base.

(*iii*) The base carries the remainder of the load.

1·11. Stresses in Chimney Plates

In determining the stresses it is assumed that the chimney as a vertical beam and the bending stress is calculated from the simple theory of bending

$$f_b = \frac{My}{I}$$

Let the external diameter $= D$

Internal diameter $= D_1$

Thickness of chimney wall=t

Then the moment of inertia$=\frac{\pi}{64}(D^4-D^4{}_1)$

$$y=\frac{D}{2} \text{ and } D-D_1=2\,t$$

$$I=\frac{\pi}{64}\times D^4-(D-2t)^4$$

$$f_b=\frac{32\,MD}{\pi\,[D^4-(D-2\,t)^4]}$$

$$=\frac{32\,MD}{\pi\,(D^4-D^4+8D^3\,t-24\,D^2t+32\,Dt^3-16t^4)}$$

$$=\frac{32\,MD}{\pi\,(8\,D^3t-24\,D^2t^2+32\,Dt^3-16\,t^4)}$$

As the values of t^2, t^3 and t^4 are quite small

$$f_b=\frac{32\,MD}{8\,\pi\,D^2t}=\frac{4\,M}{\pi\,D^2t}$$

The thickness of the plate required

$$t=\frac{4\,M}{\pi\,D^2f_b}$$

1·12. Permissible Stresses

In tension f_t=1200 kg/cm²

In compression $f_c=1200\left(1-{\cdot}00075\,\frac{D}{t}\right)$

2. Design of a Guyed Steel Unlined Chimney

2·1. Data

Diameter of the chimney $= D = 2$ m
Height $= H = 30$ m
Wind load as per I.S-875
Materials; Steel plate widths available $= 1\cdot5$ m and 2 m
Laps $= 15$ cm
Shape factor for circular section $= 0\cdot7$
Location zone -2

2·2. Permissible Stresses

In tension $= 1200$ kg/cm^2

$$\text{In compression} = 1200\left(1 - 0\;00075\,\frac{D}{t}\right)$$

Efficiency of joints in compression $= 90\%$
in tension $= 70\%$

2·3. Loads

Due to (*i*) self weight of chimney (*ii*) horizontal wind pressure upto 30 m height in zone $-2 = 150$ kg/m^2 (*iii*) Vertical component of guy wires shall be considered in the design.

2·4. Self Weight

Using 6 mm plates and 15% extra for laps
Self weight per metre height $= \pi \times D \times t \times 7850 \times 1\cdot15$

$$= \pi \times 2 \times \frac{6}{10 \times 100} \times 7850 \times 1\cdot15 = 342 \text{ kg}$$

2·5. Arrangement of Guys

Use one set of 4 guy wires at $\frac{H}{3}$ from top

$$\text{Location of collar, therefore} = \frac{30}{3} = 10 \text{ m}$$

16-mm diameter galvanized stranded guy wries with initial tension of 350 kg/cm^2 with an angle of 45° are used.

2·6. Stresses

(*i*) due to self weight
Total self weight of chimney $= W = 342 \times 30 = 10{,}260$ kg

Area of cross section of the chimney

$$A=\frac{\pi}{4}(D^2-D^2{}_1)=\frac{\pi}{4}(200^2-198{\cdot}8^2)$$

$$A=376 \text{ cm}^2$$

Approximate area for thin steel stack

$$=\pi\times\text{average diameter}\times\text{thickness}=\pi\times\frac{398{\cdot}8}{2}\times 0{\cdot}6=373 \text{ cm}^2$$

Compressive stress due to self weight per cm length of circumference $=\frac{W}{\pi D}=\frac{10260}{\pi\times 200}=16{\cdot}4$ kg/cm

(*ii*) due to vertical component in the guy wire produced by wind load

Total wind load $=P_w=0\ 7\times 150\times 30\times 2=6300$ kg

Pull on the guy wires

The guy wires must be designed to take care of the entire wind reaction at a collar. The maximum pull in any guy wire will occur when the wind blows along that guy wire.

$$R_c=\text{Reaction at collar level}=\frac{P_w H}{2\,h_1}=\frac{P_w H\times 3}{2\times 2\,H}=\tfrac{3}{4}P_w$$

$$=\tfrac{3}{4}\times 6300=4725 \text{ kg}$$

$$P_g=R_c \operatorname{cosec}\theta=4725\times\operatorname{cosec}45^\circ=4725\times 1{\cdot}414=6575 \text{ kg}$$

Total pull on the guy wire $=R_c \operatorname{cosec}\theta+$pull due to initial tension

$$=6675+350\times\frac{\pi}{4}\times(1{\cdot}6)^2=6675+705=7380 \text{ kg}$$

Stress in guy $=\frac{7380\times 4}{(1{\cdot}6)^2}=3670$ kg/cm²

Total maximum vertical component of the pulls of the guys at collar level

$$P_v=R_c \cot 45^\circ+4\times 705\cos 45^\circ$$

$$=4725\times 1+\frac{4\times 705}{1{\cdot}414}=4725+2000=6725 \text{ kg}$$

Compressive stress due to vertical component

P_v per cm length of circumference $=\frac{P_v}{\pi D}=\frac{6725}{\pi\times 200}=10{\cdot}7$ kg/cm

(*ii*) Due to horizontal wind pressure

Maximum bending moment due to wind pressure at collar level

$$=\frac{P_w\ H}{18}$$

All sections are designed to resist this bending moment

$$M_c=\frac{6300\times 30}{18}=10{,}500 \text{ m kg}$$

Maximum compression stress per cm length of circumference

$$=\frac{4\ M_c}{\pi\ D^2}=\frac{4\times10500\times100}{\pi\times200\times200}=33{\cdot}5\ \text{kg/cm}$$

Maximum compressive stress per cm length of circumference due to all loads$=16{\cdot}4+10{\cdot}7+33{\cdot}5=60{\cdot}6$ kg/cm

Maximum tensile stress per cm length of circumference $=16{\cdot}4+10{\cdot}7-33{\cdot}5=-6{\cdot}4$ kg/cm

2·7. Number of Sections

Using 1·5 m wide plates with laps of 15 cm for joints

$$\text{Number of plates required}=\frac{30}{1{\cdot}35}=22\ \text{plates}$$

2·8. Joints

Using power driven field rivets

Strength of rivets in single shear (increasing the allowable stress by 25%)$=\pi\frac{(1{\cdot}75)^2}{4}\times945\times1{\cdot}25=2840$ kg

Strength of rivet in bearing$=f_b\times d\times t=2125\times1{\cdot}75\times{\cdot}6\times1{\cdot}25$ $=2790$ kg

$$\text{Maximum shear}=S=\frac{5}{12}P_w=\frac{5}{12}\times6300=2620\ \text{kg}$$

$$\text{Shear per cm length of circumference}=\frac{2620}{\pi\times200}=4{\cdot}19\ \text{kg/cm}$$

This is negligible

However use 16 mn ϕ rivets on 6 mm plates with a pitch of 6 cm c/c in the vertical and horizontal seams.

2·9. Shear Stress

Maximum shear$=S=2620$ kg

$$\text{Shear stress}=\frac{S}{A}=\frac{2620}{376}=7\ \text{kg/cm}^2$$

Therefore the stack is safe in compression, tension and in shear.

2·10. Band at Top

A flat steel band of 12 mm$\times$60 mm wide should be riveted to the top plate of the stack to stiffen the free end or angle of ISA 60$\times$60$\times$10 mm may be used.

2·10·1. Bottom Band

The bottom band of the stack should be angle ISA 10065 with 10 mm thick with longer leg-horizontal

2·11. Foundation Bolts

At least 20 mm ϕ—12 numbers of bolts shall be used to connect the angle to the base.

2·12. Base Plate

When the chimney rests on its foundation and the breech

opening is above the base, 10 mm thick circular plate should be placed between the bottom band and foundation. Foundation bolts connecting the bottom band angle, circular plate and foundation block, shall be at least 30 cm c/c

$$\text{Number of bolts required} = \frac{\pi \times 200\ \phi}{30\ \phi} = 21 \text{ nos}$$

Minimum diameter of bolt = 20 mm

2·13. Clean Out Door

If the stack is supported in such a manner that the gasses enter through the base, then there is no need for a cleanout door

2·14. Breech Opening

If the chimney rests on a foundation and the breech opening is above the base, then a clean out door of size 50 cm × 75 cm should be placed at the bottom of the stack. The opening should be stiffened by ISA 50×50×6 angle all around the opening

2·15. Collars

The collar band should be 75 mm × 16 mm thick plate bolted to the chimney by 22 mm dia 4-eye bolts symmetrically placed for attaching the guy wires

2·16. Guy Wires

16 mm diameter galvanized cables should be used with 20 mm turn buckles.

3. Self Supporting Steel Chimneys General Information

3·1. Introduction

Self supporting chimneys are usually cylindrical in section with flared conical section at the bottom. Usual size is around 10 m in diameter 100 m in height. These are usually lined over the entire height.

3·2. Self Weight

The self weight is calculated taking steel weight as 7850 kg/m^3 and adding 20% for laps. The weight of lining is usually taken as 2000 kg/m^3 for the fire brick chimney.

3·3. Conical Section at Base

The function of the conical section at the base is to reduce the stresses in steel in the base of the chimney. It also transfers the loads to the foundation. Usually the height of this section is a sixth of the height of the chimney and its diameter is 1·5 to 1·75 times the diameter of the chimney.

3·4. Design

The self supporting steel chimneys are treated as vertical cantilever beams subjected to loads caused by the horizontal wind pressure. The horizontal joints are designed by providing sufficient number of rivets to transmit a minimum compressive stress occuring at any part of the joint.

3·4·1. Plates

Minimum thickness of plates shall not be less than 6 mm.

Minimum thickness of top two layers of plates shall not be less than 8 mm width of plates 1·75 m to 2 m.

3·4·2. Foundation Bolts and Stresses

Approximate tensile stress due to wind moment $= \dfrac{4M_w}{\pi D^2_b}$

where M_w = bending moment due to wind at base

D_b = diameter of chimney at the base.

Compressive stress due to self weight of the chimney $= \dfrac{W_s}{\pi D_b}$

where. W_s = self weight of the chimney

Maximum tensile stress per unit length of circumference

$$=\frac{4\,M_w}{\pi\,D^2_b}-\frac{W_s}{\pi\,D_b}$$

If number of bolts resisting this tension is N, each bolt resists a length $=\frac{\pi\,D_b}{N}$

$$\text{Maximum tensile stress per bolt}=\frac{\pi\,D_b}{N}\left(\frac{4\,M_w}{\pi\,D^2_b}-\frac{W_s}{\pi\,D_b}\right)$$

$$=\frac{4\,M_w}{N\,D_b}-\frac{W_s}{N}$$

Let the initial tension in the bolt be Ti

$$\text{Total maximum tension in the bolt}=\frac{4\,M_w}{N\,D_b}-\frac{W_s}{N}+Ti$$

Minimum number of bolts shall not be less than eight. Normally 10 to 12 or more are used.

3·5. Base Plate

Base plates can be either cast iron, cast steel or structural steel. Allowable stress on concrete is usually taken as 40 kg/cm² and the allowable stress in bearing plate is 1890 kg/cm².

The maximum compressive stress on the base plate is caused by the weight of steel, lining and the pressure due to wind.

Minimum compressive stress per unit length of circumference

$$=\frac{4\,M_w}{\pi\,D^2_b}+\frac{W_s+W_l}{\pi\,D_b}$$

where W_l=weight of lining

$$\text{Bending stress}=f_b=\frac{My}{I}$$

$$M=\frac{f_c\,b^2}{2}$$

$$y=\frac{t}{2},\quad I=\frac{t^3}{12}$$

$$f_b=\frac{f_c b^2\times t\times 12}{2\times 2\times t^3}=\frac{3f_c\,b^2}{t^2}$$

$$t^2=\frac{3\,f_c\,b^2}{f_b}$$

$$t=b\sqrt{\frac{3\,f_c}{f_b}}$$

$$\text{Shear stress in the base plate}=\frac{b\times f_c}{t\times 1}$$

3·6. Foundation

Depth of foundation varies from 1·5 m to 3·5 m. The foundation block is either circular or octagonal.

4. Design of a Self Supporting Chimney

4·1. Data

Height of the chimney =100 m
Diameter =5 m
Location =zone−2
Bearing capacity =20 t/m²
Depth of foundation =5 m

Lining 10 cm for top 30 m height and 20 cm for the rest of the height.

Allowable stresses

$$f_c = 1200\left(1 - \cdot 00075\,\frac{D}{t}\right)\ \text{kg/cm}^2$$

$$t_t = 1200\ \text{kg/cm}^2$$

4·2. Proportionng

Diameter =5 m

Diameter of flared conical portion =1·5 *D* to 1·75 *D* =8 m (say)

$$\text{Height of conical portion} = \frac{H}{6} = \frac{100}{6} = 16\ \text{m}$$

4·3. Breech Opening Number of plates

Breech opening =3 m×8 m

Using 2 m wide plates with 15 cm overlaps for the 80 m from top

$$\frac{84}{1 \cdot 85} = 45 \text{ plates are needed}$$

In flared portion of 16 m the overlaps may be about 25 cm
Use 8−2·25 m wide plates to make up a height of 16 m

4·4. Wind Pressure and Moments Section (a-a) 100 m—80 m

Since the chimney is loacted in zone−2 the wind pressure varies 0 to 100 m as per the code IS−875

Wind pressure at section *a—a*

M_w (*a—a*)=average wind pressure×

Shape factor×projected area of the chimney obstructi̇ng air flow×lever arm =(187×0·7×5×20) ×10=P_a×10=13050×10 =130500 m kg.

Using 6 mm plates and adding 20% extra for joints and rivets

$W_S\ (a-a) = \pi D \times t \times h \times 7850 \times 1{\cdot}2$

$= \pi \times 5\ \dfrac{6}{100 \times 100} \times 20 \times 7850 \times 1{\cdot}2$

$= 14800 \times 1{\cdot}2 = 18760$ kg

Uusing 10 cm thick fire brick lining for the top 30 m

$W_L\ (a-a) = \pi \times 5 \times 0{\cdot}1 \times 20 \times 2000 = 62{,}800$ kg

Thickness of the plate required on compression side

$$t = \frac{\dfrac{4\ M_w}{D} + W_s + W_L}{\pi D \eta\ f_c}$$

f_c for 6 mm plates $= 1200\left(1 - {\cdot}00075 \times \dfrac{D}{t}\right)$

$= 1200\left(1 - {\cdot}00075 \times \dfrac{500}{0{\cdot}6}\right) = 447{\cdot}5$ kg/cm^2

$$t = \frac{\dfrac{4 \times 13050000}{500} + 18760 + 62800}{\pi \times 400 \times 0{\cdot}9 \times 447{\cdot}5}$$

$$= \frac{102100 + 18760 + 62800}{\pi \times 500 \times 0{\cdot}9 \times 447{\cdot}5}$$

$$= \frac{182660}{\pi \times 500 \times 0{\cdot}9 \times 447{\cdot}5} = 2{\cdot}87 \text{ cm} = 2{\cdot}87 \text{ mm} < 6 \text{ mm assumed}$$

Use 8 mm plates for the top two layers to prevent corrosion etc. and 6 mm plates for the 10 layers

4·4·1. Joints

16 mm dia. field rivets—single riveted lap joints. Since the wind loads are considered, the allowable stresses in the rivets may be increased by 25%

Strength of rivet in single shear

$= \dfrac{\pi d^2}{4} fs = \dfrac{\pi}{4}(1{\cdot}75)^2 \times 1{\cdot}25 \times 945 = 2840$ kg

Strength of one rivet in bearing $= f_b \times d \times t$
$= 2125 \times 1{\cdot}25 \times 1{\cdot}75 \times 0{\cdot}6 = 2790$ kg

Force per cm length of circumference of chimney

$$= \frac{\dfrac{4M_w}{D} + W_s + W_L}{\pi D} = \frac{182660}{\pi \times 500} = 116 \text{ kg}$$

Pitch of the rivets $= \dfrac{2790}{115} = 116$ kg

Minimum pitch $= 2{\cdot}5 \times$ dia. of hole $= 2{\cdot}5 \times 1{\cdot}75 = 4{\cdot}4$ cm

Minimum pitch $= 12 \times$ thickness of plate $= 12 \times {\cdot}6 = 7{\cdot}2$ cm

Therefore use 16 mm dia. rivets rivets at 6 cm c/c single riveted lap joints in the horizontal seams as well as in the vertical seams.

4 4·2. Check for the Thickness of the Plate on Tension Side

Efficiency of the joint on tension side

$$=\frac{6-1{\cdot}75}{6}\times 100=71\%$$

t=thickness of the plate required with this efficiency.

$$=\frac{\frac{4M_w}{D}-W_s}{\pi D\eta f_t}=\frac{\frac{4\times 13050000}{500}-18760}{\pi\times 500\times 0{\cdot}71\times 1200}$$

$$=\frac{102{,}100-18960}{\pi\times 500\times 0{\cdot}71\times 1200}=\frac{83340}{\pi\times 500\times 0{\cdot}71\times 1200}$$

$=0{\cdot}0621$ cm$=\cdot 62$ mm<6 mm used.

4·5. Section (b—b) 80 m—70 m

Wind moment at sectiou $(b-b)$

$M_w\,(b-b)=P_a\times 20+P_b\times 5=13050\times 20+(0\;7\times 10\times 5\times 180)5$

$=261{,}000+6300\times 5=261{,}000+31500=292{,}500$ m kg

Try 8 mm plates between the sections $(a-a)$ and $(b-b)$

$W_s\,(b-b)=18760+\pi D\text{th}\times 7850\times 1{\cdot}2$ (20% extra for laps)

$$=18760+\pi\times 5\;\frac{8}{10\times 100}\times 10\times 7850\times 1{\cdot}2=18760+11850$$

$=30{,}610$ kg

$W_L\;(b-b)=62800+\pi\times 5\times 0{\cdot}1\times 10\times 2000=62800+31400$

$=94200$ kg

Thickness of the plate required on compression side

$$t=\frac{\frac{4M_w\,(b-b)}{D}+W_s+W_L}{\pi D\eta f_c}$$

$$f_c=1200(1-\cdot 00075\times\frac{500}{\cdot 8}=636\ \text{kg/cm}^2$$

Taking $\eta=90\%$ in compression

$$t=\frac{\frac{5\times 29250{,}000}{500}+30610+94200}{\pi\times 500\times\cdot 9\times 636}$$

$$=\frac{232000+30610+94200}{\pi\times 500\times 0{\cdot}9\times 636}=\frac{356810}{\pi\times 500\times 0{\cdot}9\times 636}=\cdot 396$$

$=3{\cdot}96$ mm$<$ 8mm assumed.

Use 8 mm plates from section $(a-a)$ to $(b-b)$

4·5 1. Joints

Try 20 mm dia. rivets

$$\text{Strength in single shear }=\frac{\pi d^2\times f_s}{4}$$

$$=\frac{\pi\times(2{\cdot}15)^2}{4}\times 1{\cdot}25\times 945=4270\text{ kg}$$

Strength of rivet in bearing$=f_b$ dt$=2125\times1{\cdot}25\times2{\cdot}15\times0{\cdot}8$
$=4550$ kg

Force per cm length of circumference of the chimney

$$=\frac{\dfrac{4M_w}{4}+W_s+W_L}{\pi D}=\frac{356810}{\pi\times500}=227 \text{ kg}$$

Pitch of the rivets $=\dfrac{4270}{227}=18{\cdot}8$ cm

Maximum pitch$=12\times t=12\times{\cdot}8=9{\cdot}6$ cm

Minimum pitch$=2{\cdot}5\ d=2{\cdot}5\times2{\cdot}15=5{\cdot}37$ cm

Therefore use 20 mm dia. rivets at 8 cm c/c single riveted lap joints for both vertical and horizontal seams.

4·6. Section (c—c) 70m—60m

M_w (c—c)$=13050\times30+6300\times15+({\cdot}7\times10\times5\times171)5$
$=391500+94500+5980\times5$
$=391{,}500+94500+29900=515{,}900$ m kg

Using 8 mm plates

$$W_s(c-c)=30610+\pi\times\frac{5\times8}{10\times100}\times10\times7850\times1{\cdot}2=30{,}610+11850$$
$$=42460 \text{ kg}$$

Since 20 cm thick fire brick lining starts from this level
W_L $(c-c)=94200+\pi\times5\times0{\cdot}2\times2000\times10=94200+62{,}800$
$=157{,}000$ kg

Thickness of the plate required on compression

$$t=\frac{\dfrac{4\times51590000}{500}+42460+157000}{\pi D\eta f_c}$$

$$=\frac{411000+42460+157000}{\pi\times500\times0{\cdot}9\times636}=0{\cdot}68 \text{ cm}=6{\cdot}8 \text{ mm}<8 \text{ mm}$$

4·6·1. Joints

Try single riveted lap joints

Force per cm length of circumference of the chimney

$$=\frac{411000+42460+157000}{\pi\times500}=\frac{610460}{\pi\times500}=390 \text{ kg/cm}^2$$

Pitch of 20 mm dia. rivets$=\dfrac{4270}{390}=11$ cm

Use 20 mm ϕ at 8 cm c/c

4·7. Sectiou (d—d) 60m—50m

M_w $(d-d)=13050\times40+630\times25+5950\times15$
$+(0{\cdot}7\times10\times5\times169{\cdot}5)5$
$=522000+153{,}500+81900+5830\times5$
$=522000+153{,}500+81900+29150=786{,}550$ m kg

Try 10 mm plates

$$W_s\,(d-d)=42460+\pi\times 5\ \frac{10}{10\times 100}\times 10\times 7850\times 1.2$$
$$=42460+14800=57260 \text{ kg}$$
$$W_L\,(d-d)=157000+62800=219{,}800 \text{ kg}$$

f_c for 10 mm plate $=1200\left(1-{\cdot}00075\times\frac{500}{1}\right)=750$ kg/cm²

Thickness of the plate required

$$t=\frac{\frac{4\times 78655000}{500}+57260+219800}{\pi\times 500\times 0{\cdot}9\times 750}$$
$$=\frac{630{\cdot}000+57260+219800}{\pi\times 500\times 0{\cdot}9\times 750}=\frac{907000}{\pi\times 500\times 0{\cdot}9\times 750}$$
$$=0{\cdot}855 \text{ cm}=8{\cdot}55 \text{ mm}$$

Use 10 mm plates

Try single riveted lap joints with 20 mm ϕ rivets

4·7·1. Joints

Force per cm length of circumference of the chimney

$$=\frac{907060}{\pi\times 500}=579 \text{ kg/cm}^2$$

Pitch of the rivets $=\frac{4270}{579}=7{\cdot}4$ cm c/c

Use 20 mm ϕ at 6 cm c/c

4·8. Section (e—e) 50 m—45 m

$$M_w\,(e-e)=13050\times 45+6300\times 30+5980\times 20$$
$$+5830\times 10+(0{\cdot}7\times 5\times 5\times 165)2{\cdot}5$$
$$=591{,}000+189{,}000+119600+58300+2890\times 2{\cdot}5$$
$$=591000+189000+119600+58300+7250$$
$$=906{,}850 \text{ m kg}$$

Try 12 — mm plates

$$W_s\,(e-e)=57260\ \pi\times\frac{5\times 12}{10\times 1000}\times 5\times 7850\times 1{\cdot}2=57260+8850$$
$$=66110 \text{ kg.}$$
$$W_L\,(e-e)=219800+\pi\times 5\times 0{\cdot}2\times 2000\times 5$$
$$=219800+31400=251200 \text{ kg}$$
$$f_c=1200\left(1-{\cdot}00075\times\frac{500}{1{\cdot}2}\right)=825 \text{ kg/cm}^2$$

Thickness of the plate required

$$t=\frac{\frac{4\times 9068500}{500}+66110+251200}{\pi\times 500\times 0{\cdot}9\times 825}$$
$$=\frac{722000+66110+251200}{\pi\times 500\times 0{\cdot}9\times 825}=\frac{1039310}{\pi\times 500\times 0{\cdot}9\times 825}$$
$$=0{\cdot}89 \text{ cm}=8{\cdot}9 \text{ mm}$$

Assumed plate thickness of 12 mm is too thick
Try 10 mm plates
M_W (e−e)=906850 m kg
W_S (e−e)=57260+7400=64660 kg
W_L (e−e)=251,200 kg

$$t=\frac{\frac{4\times 90685000}{500}+64660+251200}{\pi\times 500\times 0{\cdot}9\times 750}$$

$$=\frac{1037860}{\pi\times 500\times 0{\cdot}9\times 750}={\cdot}975\text{ cm}=9{\cdot}75\text{ mm}$$

Use 10 mm plates

4·8·1. Joints

Force per cm length of circumference$=\frac{1037860}{\pi\times 500}=660$ kg

Pitch of 20 mm ϕ rivets$=\frac{4270}{660}=6{\cdot}5$ cm

Use 20 mm ϕ at 6 cm c/c

4·9. Section (f—f) 45 m—40 m

Single riveted Lap joints in both vertical and horizontal seams

$$M_W\ (f-f)=13080\times 50+6300\times 35+5980\times 25+5830\times 15$$
$$+2890\times 7{\cdot}5+(0{\cdot}7\times 5\times 5\times 160{\cdot}5)2{\cdot}5$$
$$=519000+221000+149000+87900+21600+7050$$
$$=1005{,}550\text{ m kg}$$

Try 12 mm plates

$$W_S\ (f-f)=64660+\pi\times 5\times\frac{12}{10\times 100}\times 5\times 7850\times 1{\cdot}2$$
$$=64660+8850=73510\text{ kg}$$

W_L (f−f)=215200+31400=246600 kg

Thickness of the plate required

$$t=\frac{\frac{4\times 1005{,}55000}{500}+73510+282600}{\pi\times 500\times 0{\cdot}9\times 825}$$

$$=\frac{1161{,}110}{\pi\times 500\times 0{\cdot}9\times 825}=1\text{ cm}=10\text{ mm}$$

Use 12 mm plates

4·9·1. Joints

Try double riveted Lap joints with 20 mm ϕ rivets
Least strength in shear (for two rivets in single shear)

$$=\frac{\pi\times(2{\cdot}15)^2}{4}\times 1{\cdot}25\times 945\times 2=8540\text{ kg}$$

Least strength in bearing$=f_b\, d\times t\times 2$
$=2\times 2125\times 1.25\times 2{\cdot}15\times 1{\cdot}2=13700$ kg

Force per cm length of circumference$=\frac{1161{,}110}{\pi\times 500}=745$ kg

Pitch of the rivets $= \dfrac{8540}{745} = 11{\cdot}4$ cm c/c

Use 20 mm ϕ rivets at 8 cm c/c double riveted lap joints for horizontal seams and single riveted lap joints in the vertical seams.

4·10. Section (g-g) 40 m — 35 m

$$M_W\ (g-g) = 13050\times55+6300\times40+5980\times30+5830\times20 +2890\times12{\cdot}5+2800\times7{\cdot}5+(0{\cdot}7\times5\times5\times157)\times2{\cdot}5$$
$$=718000+252000+179000+116600+36000 +21000+2740\times2{\cdot}5$$
$$=718000+252000+179000+116600+36000 +21000+6850=1329450 \text{ m kg}$$

Try 12 mm plates

$W_S\ (g-g) = 73510+8850=82360$ kg

$W_L\ (g-g) = 282{,}600+31400=214000$ kg

Thickness of plate required

$$t = \frac{\dfrac{4\times132945000}{500}+82360+314000}{\pi\times500\times0{\cdot}9\times825}$$
$$= \frac{1031000+82360+314000}{\pi\times500\times0{\cdot}9\times825} = \frac{1427360}{\pi\times500\times0{\cdot}9\times825}$$
$$=1{\cdot}23 \text{ cm}=12{\cdot}3 \text{ mm}$$

Hence not sufficient

Try 14 mm plates

$$W_S\ (g-g) = 73150+\pi\times D\times t\times7850\times5\times1{\cdot}2$$
$$= 73150+\pi\times5\times\frac{14\times7850\times5\times1.2}{10\times100}$$
$$=73150+10400=83910 \text{ kg}$$
$$f_c = 1200\left(1-{\cdot}00075\times\frac{500}{1{\cdot}4}\right)$$
$$=940 \text{ kg/cm}^2$$
$$t = \frac{\dfrac{4\times132945000}{500}+83910+314000}{\pi\times500\times0{\cdot}9\times940}$$
$$= \frac{1062000+83910+314000}{\pi\times500\times0{\cdot}9\times940}$$
$$= \frac{1459910}{\pi\times500\times0{\cdot}9\times940} = 1{\cdot}09 \text{ cm}=10{\cdot}9 \text{ mm}$$

Use 14 mm plates

4·10·1. Joints

Force per cm length $= \dfrac{145910}{\pi\times500} = 960$ kg

Try double riveted lap joints using 22 mm ϕ rivets

Least strength in shear for two rivets in single shear

$= 2 \times \frac{\pi(2\cdot35)^2}{4} \times 1\cdot25 \times 945 = 10200$ kg

In bearing$= f_b d\ t \times 2 = 2125 \times 1\cdot25 \times 2\cdot35 \times 1\cdot4 \times 2 = 17500$ kg

Pitch of the rivets$= \frac{10200}{960} = 10\cdot6$ cm c/c

Use 22 mm ϕ at 8 cm c/c. Double riveted Lap joints for horizontal seems and single riveted for the verticle seams.

4·11. Section (h—h) 35m—30m

$M_w(h-h) = 13050 \times 60 + 6300 \times 45 + 5980 \times 35$
$+ 5830 \times 25 + 2890 \times 17\cdot5 + 2800 \times 12\cdot5$
$+ 2740 \times 7\cdot5 + (0\cdot7 \times 5 \times 5 \times 153)\ 25$
$= 783000 + 284000 + 209000 + 146100 + 50{,}500$
$+ 35000 + 20500 + 2670 \times 2\cdot5$
$= 783000 + 284000 + 209000 + 146100 + 50500$
$+ 35000 + 20500 + 6750 = 1534850$ m kg

Try 14 mm plates

$W_s\ (h-h) = 83910 + 10400 = 94310$ kg

$W_L\ (h-h) = 314000 + 31400 = 345400$ kg

$$t = \frac{\frac{4 \times 153485000}{500} + 94310 + 345400}{\pi \times 500 \times 0\cdot9 \times 940}$$

$$= \frac{1561710}{\pi \times 500 \times 0\cdot9 \times 940} = 1\cdot1 \text{ cm} = 11 \text{ cm}$$

Use 14 mm plates

4·11·1. Joints

Force per cm length of circumference

$= \frac{1561710}{\pi \times 500} = 1000$ kg

Pitch of 22 mm ϕ using double riveted lap joints in the horizontal seams$= \frac{10200}{1000} = 10\cdot2$ cm c/c

Use 22 mm ϕ at **6** em c/c

Use single riveted lap joints In the vertical seams

4·12. Section (k—k) 30m—16m

$M_w\ (k-k) = 13050 \times 74 + 6300 \times 59 + 5980 \times 49$
$+ 5830 \times 39 + 2890 \times 31\cdot5 + 2800 \times 26\cdot5$
$+ 2740 \times 21\cdot5 + 2670 \times 16\cdot5 + (0\cdot7 \times 5 \times 16 \times 150)\ 7$
$= 965000 + 372000 + 292000 + 229000 + 91000 + 74200$
$+ 59000 + 44200 + 8400 \times 7 = 2184200$ m kg

$W_s\ (k-k) = 94310 + \pi\ Dt \times 7850 \times 14 \times 1\cdot2$

$= 94310 + \pi \times 5 \times \frac{1\cdot8}{100} \times 7850 \times 14 \times 1\cdot2$

$= 94310 + 37320 = 131630$ kg

$$W_L\ (h-h)=345400+5\times\frac{20}{100}\times 2000\times 14=345400+88000$$

$$=433400 \text{ kg}$$

$$f_c=1200\left(1-{\cdot}00075\times\frac{500}{1{\cdot}8}\right)=955 \text{ kg/cm}^2$$

$$t=\frac{\dfrac{4\times 2184200}{500}+131630+433400}{\pi D\eta\times f_c}$$

$$=\frac{1700000+131630+433400}{\pi\times 500\times 0{\cdot}9\times 955}=\frac{2265030}{\pi\times 500\times 0{\cdot}9\times 955}$$

$$=1{\cdot}68 \text{ cm}$$

Use 18 mm plates

4·12.1. Joints

Try double riveted lap joints using 25 mm ϕ

Least strength in single shear for two rivets

$$=2\times\frac{\pi(2{\cdot}65)^2}{4}\times 1{\cdot}25\times 945=13000 \text{ kg}$$

In bearing $=2\times f_b\times d\times t=2\times 1{\cdot}25\times 2125\times 2{\cdot}65\times 1{\cdot}8=12600$ kg

Force per cm length of circumference $=\dfrac{2265030}{\pi\times 500}=1440$ kg

Pitch $=\dfrac{12600}{1440}=8{\cdot}75$ cm

Use double riveted lap joints with 25 cm ϕ at 8 cm c/c in the horizontal seams and single riveted lap joints for the vertical seams.

4·13. Section (L—L) 16m—0m (Flared portion)

$$M_w\ (L-L)=13050\times 90+6300\times 75+5980\times 65+5830\times 55$$
$$+2890\times 47{\cdot}5+2800\times 42{\cdot}5+2740\times 37{\cdot}5+2640\times 23{\cdot}5$$
$$+8400\times 2300+\left[0{\cdot}7\times\left(\frac{5+8}{2}\right)\times 16\times 150\right]8$$
$$=1179000+472000+389000+322000+136500+119000$$
$$+103000+86000+193000+10900\times 8=3087000 \text{ m kg}$$

Try 20 mm thick plates

$$W_s\ (L-L)=131630+\pi \text{ dt }\times 7850\times 16\times 1{\cdot}2$$
$$=131630+\left(\frac{5+8}{2}\right)\times\frac{2}{100}\times 7850\times 16\times 1{\cdot}2$$
$$=131630+61200=192830 \text{ kg}$$

$$W_L\ (L-L)=433400+\pi\left(\frac{8+5}{2}\right)\times\frac{20}{100}\times 2000\times 16$$
$$=433400+130000=563400 \text{ kg}$$

Thickness of the plate required

$$t=\frac{\dfrac{4\times 3087000\times 100}{500}+192837+563400}{\pi D\eta f_c}$$

f_c for 20 mm plates (average diameter 6·5 m)

$$f_c=1200\left(1-{\cdot}00075\frac{D}{t}\right)=1200\left(1-{\cdot}00075\times\frac{650}{2}\right)$$

$=905\ \text{kg/cm}^2$

$$t=\frac{\dfrac{4\times3087000\times100}{500}+192830+563400}{\pi\times650\times{\cdot}9\times905}$$

$$=\frac{2460{,}000+192830+563400}{\pi\times650\times0{\cdot}9\times905}=\frac{3216230}{\pi\times650\times0{\cdot}9\times905}$$

$=1{\cdot}94\ \text{cm}=19{\cdot}4\ \text{mm}$

Use 25 mm thick plates

4·13·1. Joints

$$\text{Force per cm length of circumference}=\frac{3216230}{\pi\times650}=1580\ \text{kg}$$

$$\text{Pitch of 25 mm }\phi\text{ rivets}=\frac{12600}{1580}=8\ \text{cm}$$

(double riveted lap joint)

Use 25 mm ϕ rivets at 7 cm c/c (2·5 d)

4·14. Foundation Bolts

Minimum number of bolts shall be not less than 8

Try 48 bolts

$$T_{max}=\frac{4M_w}{N\,D_b}-\frac{W_s}{N}+T_i$$

Where T_i is the initial tension in the bolt$=300$ to $400\ \text{kg/cm}^2$

$$T_{max}=\frac{4\times3087000}{48\times8}-\frac{192830}{48}+300=32100-4020+300$$

$=28380\ \text{kg}$

Net area of bolt required

$$=\frac{T_{max}}{\text{allowable stress}}=\frac{28380}{1{\cdot}25\times1260}=18\ \text{cm}^2$$

$$\text{Diameter of the bolt}=d=\sqrt{\frac{18\times4}{\pi}}=4{\cdot}79\ \text{cm}$$

Use 50 mm dia. bolts

4·15. Base Plate

Maximum compressive force at the base per meter length circumference of the chimney

$$=\frac{4M}{\pi D_b^2}+\frac{W_s+W_L}{\pi D_b}=\frac{4\times3087000}{\pi\times8\times8}+\frac{192830+563400}{\pi\times8}$$

$$=61100+\frac{756230}{\pi+8}=61100+30100=91200\ \text{kg}$$

Allowable stress on concrete base $=1{\cdot}33\times40=53{\cdot}33\ \text{kg/cm}^2$

Force per cm length of circumference $= \frac{91200}{100} = 912$ kg

Width of base plate required $= \frac{912}{53\cdot33} = 17\cdot3$cm

Use 20 cm wide plate

4·15·1. Base Plate Thickness

Use 2 angles of ISA$-100\times100\times10$ and 25 mm ϕ rivets to connect the chimney plate to the base plate

b=projection$=10-2=8$ cm

Pressure below the base plate $= \frac{912}{20\times1} = 45\cdot6$ kg/cm^2

Bending moment at xx $= \frac{45\cdot6\times8\times8}{2}$

$$M = 1460 \text{ cm kg}$$

$$f_b = \text{bending stress} = \frac{M}{Z} = \frac{M}{\frac{1}{6}t^2} = \frac{6M}{t^2}$$

$$t = \sqrt{\frac{6M}{f_b}} = \sqrt{\frac{6\times1460}{1890\times1\cdot33}}$$

$$= \sqrt{3\cdot48} = 1\cdot86 \text{ cm}$$

Use 20 mm thick plate.

Shear stress in the plate

$$= \frac{f_c b}{t\times1} = \frac{45\cdot6\times4}{1\times1} = 182\cdot4 \text{ kg/cm}^2 = 1100 \text{ kg/cm}^2$$

4·16. Foundation Block

Wind moment about m$-$m

$$M_f = 13050\times95+6300\times80+5980\times70+5890\times60+2890\times52\cdot5$$
$$+2800\times47\cdot5+2740+42\cdot5+2640\times37\cdot5$$
$$+8400\times28+10900\times13$$
$$=1242000+504000+418600+353400+151000$$
$$+133000+116000+99000+236000+142000$$
$$=3395000 \text{ m kg} = 3395 \text{ mt}$$

Weight of foundation block

$$W_f = \frac{\pi h}{12}(D_1^2+D_2^2+D_1D_2)\times2\cdot4 \text{ t}$$

$$= \frac{\pi\times5}{12}(8^2+18^2+18\times8)\times2\cdot4 = 1670\ t$$

4·16·1. Stresses Case (i) no Lining

$$\text{Minimum stress} = \frac{W_s+W_f}{A_f} - \frac{M_f D_f}{2I_f}$$

$$= \frac{(192\cdot83+1670)}{\frac{\pi\times18^2}{4}} - \frac{3395\times18}{\frac{2\times\pi\times18^4}{64}}$$

$$=\frac{1862{\cdot}83}{255}-\frac{61000}{10300}=7{\cdot}35-5{\cdot}9$$

$=1{\cdot}45$ t/m^2 (compressive)

4·16·2. Case (ii) When Lining is Present

Maximum stress $=13{\cdot}25$ t/m^2 <20 t/m^2

$$\text{Maximum stress}=\frac{192{\cdot}83+563{\cdot}4+1670}{\frac{\pi\times 18^2}{4}}+5{\cdot}9$$

$$=\frac{2426{\cdot}23}{255}+5{\cdot}9=9{\cdot}5+5{\cdot}9$$

$=15{\cdot}4$ t/m^2 (compressive) <20t/m^2

Minimum stress $=9{\cdot}5-5{\cdot}9=3{\cdot}6$t/m^2 (comp)

Section	Elevatian from bottom m	Basic wind preesure zone-2 kg/m²	Average wind pressure at the section kg/m²	Wind moment at the level of of the sectlon M_w m kg	Self weight of chimney plates W_s kg	Weight of lining W_L kg	Thickness of the plate t mm	Joint details: Horizontal seams	Joint details: Vertical seams
Top	100	191		—					
			187						
a—a	80	183		130500	18760	62800	6 mm	Single riveted Lap-16 mmφ @ 6 cm c/c	
			180						
b—b	70	177		292500	30610	94200	8 mm	S.R. Lap—20 mm φ @ 8 cm c/c	
			174·5						
c—c	60	172		515900	42460	157000	8 mm	,,	,,
			169·5						
d—d	50	167		786550	57260	219800	10 mm	S·R. Lap 20 mm φ —6 mm c/c	
			165						
e—e	45	163		906850	64660	251200	10 mm	,,	,,
			160·5						
f—f	40	158		1005550	73580	282600	12 mm	Double riveted 20 φ—8 cm c/c	S.R.—20 8 cm c/c
			157						
g—g	35	156		1329450	83910	314000	14 mm	D—R—20 mm φ 8 cm c/c	S.R.—22 —8 cm c/c
			153						
h—h	30	150		1534850	94310	345400	14 mm	D—R—22 mm φ 6 cm c/c	S.R.—22 —6 cm c/c
			150						
k—k	16	150		2184200	131630	433400	18 mm	D—R—25 mm φ 8 cm c/c	S.R.25— —8 cm c/c
			150						
l—l	00	150		3087,000	192830	563430	20 mm	D—R—25 mm φ 7 cm c/c	S.R.—25 —7 cm c/c